BORON

Synthesis, Structure, and Properties

BORON
Synthesis, Structure, and Properties

Proceedings of the Conference on Boron

Sponsored by

**Institute for Exploratory Research,
The U.S. Army Signal Research
and Development Laboratory,
Fort Monmouth, New Jersey.**

Edited by

J. A. KOHN, W. F. NYE, G. K. GAULÉ
U.S. Army Signal Research and Development Laboratory
Fort Monmouth, New Jersey

Springer Science+Business Media, LLC

1960

ISBN 978-1-4899-6265-2   ISBN 978-1-4899-6572-1 (eBook)
DOI 10.1007/978-1-4899-6572-1

Copyright © 1960 by Springer Science+Business Media New York
Originally published by Plenum Press, Inc. in 1960.
Softcover reprint of the hardcover 1st edition 1960

Library of Congress Catalog Card Number 60-13945

FOREWORD

On the sixteenth of September in 1804 a balloon rising into the cool air over Paris carried aloft a young French scientist who was-- in a very real sense--a worthy precursor of the present-day space scientist and geophysicist. Joseph Louis Gay-Lussac ascended to the then great elevation of 23,000 feet where, at a temperature of 9½ degrees below freezing, he made observations on terrestrial magnetism, measured temperature and humidity of the atmosphere, and captured samples of air for analysis.

Five years later, Gay-Lussac held the Chair of Physics at the Sorbonne and the Chair of Chemistry at the École Polytechnique. It was in this same year, 1809, that he succeeded in isolating elemental boron, a feat which was accomplished through the reduction of boric acid by potassium.

The historical setting sketched in the preceding paragraphs has multiple relevancy to the context in which the conference here reported is set. The year 1959 finds man very actively investigating his extraterrestrial environment, industriously seeking new materials with which to do so, and wisely combining his capabilities in physics and chemistry and related sciences in the furtherance of his endeavors.

The Conference on Boron was held September 18-19, 1959, at Asbury Park, New Jersey, not far from the home of the sponsoring organization, the Institute for Exploratory Research, a part of the United States Army Signal Research and Development Laboratory, Fort Monmouth.

The conference title is conveniently short, but deceptively broad; it unintentionally masks the modest scope of the conference. In fact, the deliberately circumscribed objectives and limited scope, as set forth in the conference announcements, were addressed to the solid state physics and chemistry aspects of elemental boron and closely related (high-boron) borides.

In order to keep within reasonable bounds, conferees were invited to concentrate on the following topical areas:

a. Crystallization, purification, and crystal growth.
b. Crystal structure and bonding.
c. Fundamental physical properties, especially electronic and optical.

Papers dealing with boron chemistry in general were excluded, cognizance being taken of the recent symposia of the Committee on

v

Boron Chemistry of the Division of Inorganic Chemistry of the American Chemical Society. Further, with respect to certain newer fields of interest, papers dealing with boron and borides in relation to high-energy fuels or propellants and those dealing specifically with nuclear (neutron absorber) applications were also excluded.

Within the thus delimited scope there has been in recent years a renaissance of interest in the electronic (especially semiconductor), optical (especially infrared), thermal, mechanical, and chemical bulk and surface properties of boron; correspondingly, interest in the dependence of those properties upon the compositional and structural purity of the boron has developed.

While recent research in this field has been fruitful and promising, further progress has been frustrated: first, by an incomplete understanding of the crystal structure of the most common form of elemental boron, of the relationship of the structure of boron to that of high-boron borides, of the structural and thermodynamic relationships of the different polymorphic varieties of boron, and of the conditions under which they form; secondly, by the lack of very pure boron in any form and of boron single crystals of a size adequate for research. Thus, the nature of the interatomic bonding in boron, reflected in its high melting point and substantial hardness, augurs well for the realization of unusual high-temperature properties while at the same time imposing difficult barriers both in research and in the exploitation of research findings.

It was a specific objective of the Conference Committee to lower some of these barriers to research progress by bringing together for discussion and exchange of information those engaged in research in this area of solid state science. Those who participated, by the presentation of papers and by discussion of results and interpretations, are listed on the page following this foreword. They numbered 95 persons in all, coming from 7 university and research institute laboratories, 7 government research laboratories, and 23 industrial laboratories, from many parts of the United States, and from western Europe.

In addition to the brief discussions which followed each presentation, a summary round table was held. Under the able and stimulating chairmanship of Hubbard Horn, many aspects of the purification, crystallization, structure, and properties of boron were discussed in the light of information presented and views expressed.

The conference was, of course, most fruitful to those who participated, but the Conference Committee hopes that this compilation of the papers as presented (and revised in part to reflect the sense of the discussions) may prove helpful to others and stimulate further constructive interchanges.

The Committee which initiated the conference, arranged the program, and has concerned itself with the assembly of the papers consisted of the following: G. P. Dunbar, G. K. Gaulé, J. A. Kohn, W. F. Nye, and S. Benedict Levin. Gaulé and Kohn were particularly responsible for the arrangement of the conference session programs, and they, together with J. L. Hoard and F. Hubbard Horn, served as chairmen of the sessions.

The Committee and the conferees are particularly indebted to Mrs. Dunbar and her assistants for their labors in the initial planning, in the arrangement of facilities, and in the assembly of the manuscripts.

S. BENEDICT LEVIN

CONFERENCE PARTICIPANTS

J.C. Barett	Office of the Secretary of Defense, Washington, D.C.
K.E. Bean	The Eagle-Picher Company, Miami, Okla.
W. Brenner	New York University, New York, N.Y.
R. Brungs	St. Louis University, St. Louis, Mo.
A. Cane	Hughes Tool Company, Aircraft Div., Culver City, Cal.
G.G. Collins	U.S. Borax Research Corp., Anaheim, Cal.
H.A. Eick	Michigan State University, East Lansing, Michigan
R.C. Ellis	Raytheon Co., Waltham, Mass.
K. Eriks	Boston University, Boston, Mass.
W.R. Eubank	Minnesota Mining and Manufacturing Co., St. Paul, Minn.
C. Feldman	U.S. Naval Research Laboratory, Washington 25, D.C.
W.O. Freitag	Foote Mineral Company, Berwyn, Pa.
G.E. Gazza	Watertown Arsenal Laboratories, Watertown, Mass.
E.S. Greiner	Bell Telephone Laboratories, Murray Hill, N.J.
A.K. Hagenlocher	Telefunken, Ulm, West Germany
D.M. Harris	Monsanto Chemical Co., Dayton, Ohio
J.L. Hoard	Cornell University, Ithaca, N.Y.
F.H. Horn	General Electric Co. Research Laboratory, Schenectady, N.Y.
V.P. Jacobsmeyer, Jr.	St. Louis University, St. Louis, Mo.
P.D. Johnson	Atomics International, Inc., Canoga Park, Cal.
P.H. Keck	Sylvania Electric Products, Inc., Bayside, N.Y.
W. Kolyer	American Potash and Chemical Corp., New York, N.Y.
S.W. Kurnick	General Atomic, San Diego, Cal.
S. LaPlaca	Polytechnic Institute of Brooklyn, Brooklyn, N.Y.
R.F. Lever	Philco Corp., Philadelphia, Pa.
C.H. Lewis	Metal Hydrides, Inc., Beverly, Mass.
W.E. Medcalf	The Eagle-Picher Company, Miami, Okla.

M. Miksic   Polytechnic Institute of Brooklyn,
                            Brooklyn, N.Y.
G.T. Miller   Hooker Chemical Corporation,
                            Niagara Falls, N.Y.
L.A. Murray   International Telephone and Telegraph
                            Laboratories, Nutley, N.J.
N.P. Nies   U.S. Borax Research Corp., Anaheim, Cal.
C.A. Nixon   U.S. Army Rocket and Guided Missile
                            Agency, Redstone Arsenal,
                            Huntsville, Ala.
Q.D. Overman, Jr.   Experiment Incorporated, Richmond, Va.
J. Powderly   The Eagle-Picher Company, Miami, Okla.
C.F. Powell   Battelle Memorial Institute, Columbus, Ohio
H.F. Rizzo, Capt.   Wright Air Development Center,
                            Wright-Patterson Air Force Base, Ohio
R.F. Rolsten   U.S. Borax Research Corp., Anaheim, Cal.
J.W. Ross   Texas Instruments, Inc., Dallas, Texas
B. Rubin   Air Force Cambridge Research Center,
                            L.G. Hanscomb Field, Bedford, Mass.
W. Ruigh   Wright Air Development Center,
                            Wright-Patterson Air Force Base, Ohio
P. Russell   Hoffman Science Center, Los Angeles, Cal.
H. Shenker   U.S. Naval Research Laboratory,
                            Washington 25, D.C.
R.J. Starks   The Eagle-Picher Company, Miami, Okla.
D.R. Stern   American Potash and Chemical Corp.,
                            Los Angeles, Cal.
C.P. Talley   Experiment Incorporated, Richmond, Va.
H. Thurnauer   Minnesota Mining and Manufacturing Co.,
                            St. Paul, Minn.
M. Vilella   Electro-Thermal Industries,
                            Pearl River, N.Y.
G.R. Watson   Norton Company, Chippawa, Ontario, Can.
D.N. Williams   Battelle Memorial Institute, Columbus, Ohio
H.V. Winston   Hughes Aircraft Co., Culver City, Cal.
W. Zimmerman III   U.S. Naval Research Laboratory,
                            Washington 25, D.C.

From U.S. Army Signal Research and Development Laboratory,
Fort Monmouth, N.J.

I. Adams	R. Gambino	F. Kornfeil	F. M. Ryan
P. Bramhall	G. Gaulé	R. Lefker	R. Shuttleworth
J. Breslin	R. Geraghty	F. Leonhard	R. L. Smith
J. Broder	A. A. Giardini	S. B. Levin	A. Tauber
C. Cook	M. Green	R. V. McKnight	L. Toman
S. DiVita	I. Greenberg	J. Mellichamp	W. Townes
G. P. Dunbar	J. Gualtieri	O. Moore	J. E. Tydings
D. Eckart	M. Katz	J. N. Mrgudich	S. Ullman
M. Fink	H. H. Kedesdy	W. F. Nye	C. Whinfrey
J. Finnegan	A. Kerecman	J. Pastore	G. A. Wolff
D. Fox	J. A. Kohn	E. M. Reilley	

CONTENTS

REMARKS ON STRUCTURE AND POLYMORPHISM IN BORON

J. L. Hoard[*]

The three-dimensional framework structures of alpha-rhombohedral boron, tetragonal boron, and the boron carbide phase all contain linked B_{12} icosahedra. Half of the atoms in alpha-rhombohedral boron, the polymorph formed at lowest temperatures, are used in delta linkages between icosahedra, and the structure then satisfies the electron counting set by the essentially molecular Longuet-Higgins and Roberts bonding theory. More conventional and apparently stronger linkages to icosahedra are displayed in tetragonal boron and boron carbide. It is doubtful whether icosahedra play any essential role in the as yet undetermined very complex structure of the most important polymorph, beta-rhombohedral boron. Critical examination of reported powder data shows that (1) beta-rhombohedral boron may be confidently expected from high temperature (>1500°C) preparation with or without fusion, (2) besides the three modifications established by single crystal data there are several other claimants to recognition. The multiplication of structural variants at lower temperatures probably is dictated by kinetic rather than thermodynamic factors. It is suggested that boron prepared by deposition onto heated substrates is especially prone to form monotropes, often as nonstoichiometric borides.

Inasmuch as the subject of structure and polymorphism in boron is treated at length in two articles [1,2] scheduled for early publication,[†] a unified summary of the principal conclusions recorded therein is all that can, with propriety or profit, be set down at this time.

Structure

Success in growing pure single crystals of boron has made possible the certain identification and subsequent structural investigation of three polymorphic modifications: the (low-temperature) alpha-rhombohedral [3], tetragonal [4], and (high-temperature) beta-rhombohedral [5] polymorphs. The atomic arrangements within the alpha-rhombohedral [6] and tetragonal [7,8] forms are known, while that of the very complex beta-rhombohedral [5] polymorph is pres-

---

[*] Department of Chemistry, Cornell University, Ithaca, N. Y.
[†] Editors' note: Since published.

1

ently the subject of intensive study [9]. Also pertinent to the discussion of structure is the boron carbide phase of known structural type [10, 11]. The phase is readily prepared with substitution of boron for carbon atoms within the C_3 chains so as to extend the composition range from the theoretical $B_{12}C_3$ at least as far as B_7C [12,13,14].

The three known structural types have a common structural entity [1], namely, the group of twelve boron atoms arranged at the vertices of a nearly regular icosahedron. Probably the best estimate of size for a regular B_{12} icosahedron comes from the study of the rather complex tetragonal boron [1,8]. An edge length of 1.805 ± 0.015 A yields an icosahedron which can be inscribed within a sphere 3.43 A in diameter. Far from existing as discrete islands, the icosahedra are linked into three-dimensional networks; it is, indeed, the rather fundamental variations in the mode of connecting icosahedra, on passing from one structural arrangement to another, which ultimately command greatest interest.

Tetragonal boron [1,8], space group--$P4_2/nnm$, has $a = 8.75$, $c = 5.06$ A. The fifty atoms of the cell are distributed among four equivalent B_{12} icosahedra of required minimum symmetry, $2/m$, and two "tetrahedral" positions, $\bar{4}2m$. Each boron atom of an icosahedron forms six bonds directed toward the vertices of a pentagonal pyramid, five bonds within the icosahedron, the sixth pointing outward along a quasi-fivefold axis. Ten of the twelve external bonds from a B_{12} group are direct links to ten neighboring icosahedra, while the remaining external bonds are to the tetrahedral atoms; each of these latter forms but four short links at 1.601 ± 0.005 A, a separation corresponding [1] to a bond order of unity. The best estimate of length for a typical intericosahedral bond along the quasi-fivefold axis of a linked pair is 1.68 ± 0.03 A (bond order, about ¾), substantially smaller than the intraicosahedral value of 1.805 A (bond order, about ½).

Boron carbide [10] of nearly theoretical composition, $B_{12}C_3$, space group--$R\bar{3}m$, has $a = 5.60$ A, $\alpha = 65°18'$. The cell contains one B_{12} icosahedron and one linear C_3 chain, both of required symmetry, $3m$. Each atom of a B_{12} group, as in tetragonal boron, forms six pentagonal pyramidal bonds. Six of the external bonds are direct links to neighboring icosahedra centered at points of the rhombohedral lattice, the remaining six give connections with terminal atoms of six adjacent chains. Only approximate B−B bond lengths, 1.74-1.80 A, are known [10,11].

Alpha-rhombohedral boron [6], space group--$R\bar{3}m$, has $a = 5.06$ A, $\alpha = 58°4'$. The unit cell, containing a single B_{12} icosahedron, is formally derivable from that of boron carbide through elimination of the chains, with accompanying shrinkage mostly evident in the smaller rhombohedral angle. The direct B−B links connecting icosahedra in boron carbide, involving half of all boron atoms, are retained in alpha-rhombohedral boron; the reported [6] bond distance is 1.709 A. But, whereas in boron carbide three boron atoms of three different icosahedra are bonded to a common terminal atom of a C_3 chain, in alpha-rhombohedral boron the corresponding triad of boron atoms forms a three-center or delta bond 2.03 A in length [6]. Half of the boron atoms are used to form delta bonds connecting icosahedra. Each boron atom thus employed may alternatively be said [1] to form two weak intericosahedral links of bond order < ¼.

Beta-rhombohedral boron [5,9], space group--$R\bar{3}m$, has $a = 10.12$ A, $\alpha = 65°28'$; the cell contains 108 ± 1 atoms. The identity of space group, the similarity in α, and the approximate doubling of a, as compared with either boron carbide or alpha-rhombohedral boron, suggest a further close structural relationship. Current study of the beta-rhombohedral data does not support this hypothesis; indeed, it is doubtful whether the icosahedral group plays a prominent role in the structure of this most important form of boron.

The elegant (but approximate) molecular orbital-bonding theory developed [15] for the discrete B_{12} group of regular icosahedral symmetry accounts for the instability of the $B_{12}H_{12}$ molecule. The extension, in rather arbitrary fashion, of this essentially molecular theory to the network structures of interest gives interestingly mixed results. The delta linkages connecting icosahedra in alpha-rhombohedral boron permit rationalization of the bonding pattern so as to meet the electron counting demanded by the theory [1,6]. Provided it is formulated as $C_3^{++}B_{12}^{=}$, a boron carbide of ideal composition gives the required electron count [15]. When applied, however, either to tetragonal boron [1,15] or to boron-rich boron carbide [1], the theory calls for substantially more electrons than can be present.

Polymorphism

For the discussion of polymorphism in boron, unless otherwise specified, the general preparative method is assumed to be deposition from a vapor phase onto a heated substrate.

The alpha-rhombohedral polymorph has been prepared [2] at temperatures ranging from 750°C to nearly 1200°C; it is apparently the only form which can be crystallized below approximately 1000°C. It is thermally unstable above 1200°C [2,6], transforming irreversibly into beta-rhombohedral boron at 1500°C. The kinetically sluggish and generally complex behavior of the system in the intermediate range, 1200-1500°C, is consistent (see below) with the general conclusions educed from the critical examination of reported powder data [2].

It appears [2] that beta-rhombohedral boron is to be expected from reasonably pure, high-temperature (>1500°C) preparations with or without fusion. Between 1000 and 1500°C, however, there is an astonishing proliferation [2] of claimants to recognition as distinct polymorphs. Critical analysis eliminates four of these suggested polymorphs; there remain, nevertheless, some four or five other possibilities (all in addition to the three established forms) supported by more or less convincing powder diffraction and other analytical data [2]. (Cf. also [16].) The salient point to be noted for all preparations within the 1000-1500°C band is the following: the nature of the product apparently is dictated by the detailed character of the process, the choices of substrate and of boron source being especially significant. Indeed, the formation of tetragonal boron (1100-1300°C) and of alpha-rhombohedral boron (1000-1150°C) are cases in point [2].

It is manifestly improbable that a half dozen or more different polymorphs, each thermodynamically stable within a characteristic temperature interval, can be accommodated within the 1000-1500°C band. Rather, it would appear, the choice of framework below 1500°C is determined by kinetic factors, and most, if not all, of the polymorphs thus selected are monotropic forms of boron [1,2]. The occasional direct formation at temperatures as low as 1100°C of the very complex beta-rhombohedral phase [2] may well follow true thermodynamic stability, but, even so, is only made possible by a favorable kinetic mechanism. The circumstances suggest a non-stoichiometric boride mechanism [1,2]: namely, attack upon the substrate with initiation of a structural pattern corresponding to an interstitial boride or solid solution based upon the beta-rhombohedral framework. Continued deposition from the vapor phase at a rate too fast to be matched by the radial diffusion of substrate atoms (with, of course, no change in basic framework) can then lead to recovery of the rather pure polymorph from the outer layers.

Elaboration [1,2] of the nonstoichiometric boride mechanism suggests that it can account for the proliferation of apparent boron polymorphs below 1500°C, the lowest temperature at which, it would seem, crystallization of beta-rhombohedral boron is spontaneously initiated in the absence of any other component. The dependence of product upon the choice of substrate, the boron source, and perhaps other factors is to be expected in these circumstances.

This is not to imply that a nonstoichiometric boride mechanism is always operative. The growth mechanism of essentially pure single crystals of tetragonal boron from a matrix of different structure cannot be unambiguously classified [2] on the basis of present data. It seems improbable that alpha-rhombohedral boron, the only reported boron phase of truly simple structure [1,6], is produced by a nonstoichiometric boride mechanism. Holes in the structure are hardly large enough to accommodate the usual type of substrate atom while avoiding attack upon the presumably susceptible three-center bonds which link icosahedra. It is more reasonable to suppose that the alpha-rhombohedral framework is kinetically the uniquely feasible crystalline arrangement at low temperatures, being formed from icosahedral fragments which, judging from the chemistry of the boron hydrides, are likely to be present in the vapor phase. By virtue of the three-center intericosahedral connections, the alpha-rhombohedral structure obeys the electron counting demanded by the essentially molecular Longuet-Higgins and Roberts bonding theory [1,6]. The alpha-rhombohedral polymorph might then be character-ized as the obvious anhydrogenide of the boron hydrides.

The net effect [2] of the several reported preparations of beta-rhombohedral boron is to suggest thermodynamic stability of this phase throughout most of the temperature band, say >1100°C, in which the proliferation of reported polymorphs occurs; more spe-cifically, this includes the temperature range within which it has been possible to prepare the tetragonal polymorph. The linking of icosahedra in tetragonal boron is almost surely energetically favor-able as compared with alpha-rhombohedral boron; this latter form could still be thermodynamically favored provided it were sufficiently the higher in entropy. Should, however, the relative entropy content become controlling, it ought to favor alpha-rhombohedral boron at higher, rather than, as observed, lower temperatures. This argument suggests again that formation of the alpha-rhombohedral polymorph is a consequence of kinetic simplicity; it suggests further that the beta-rhombohedral polymorph may be the thermodynamically fa-vored phase at all temperatures (and normally low pressures).

Acknowledgment

Current structural investigations of elemental boron at Cornell University are supported in part by National Science Foundation Re-search Grant NSF-G5924.

References

1. Hoard, J. L., "Structure and Polymorphism in Elemental Boron," in "From Borax to Boranes." (Advances in Chemistry Series) (Washington: Am. Chem. Soc., 1960).
2. Hoard, J. L., and Newkirk, A. E., J. Am. Chem. Soc., 82 (1960) 70.
3. McCarty, L. V., Kasper, J. S., Horn, F. H., Decker, B. F., and Newkirk, A. E., J. Am. Chem. Soc. 80 (1958) 2592.
4. Laubengayer, A. W., Hurd, D. T., Newkirk, A. E., and Hoard, J. L., J. Am. Chem. Soc. 65 (1943) 1924.
5. Sands, D. E., and Hoard, J. L., J. Am. Chem. Soc. 79 (1957) 5582.
6. Decker, B. F., and Kasper, J. S., Acta Cryst. 12 (1959) 503.

7. Hoard, J. L., Geller, S., and Hughes, R. E., J. Am. Chem. Soc. 73 (1951) 1892.

8. Hoard, J. L., Hughes, R. E., and Sands, D. E., J. Am. Chem. Soc. 80 (1958) 4507.

9. Hoard, J. L., Sands, D. E., Weakliem, H. A., and Sullenger, D. B., Work in progress.

10. Clark, H. K., and Hoard, J.L., J. Am. Chem. Soc. 65 (1943) 2115.

11. Zhdanov, G. S., and Sevast'yanov, N. G., Compt. rend. acad. sci. U.R.S.S. 32 (1941) 432.

12. Allen, R. D., J. Am. Chem. Soc. 75 (1953) 3582.

13. Bray, P. J., Atomic Energy Commission Report NYO-7624 (1958).

14. Glaser, F. W., Moskowitz, D., and Post, B., J. Appl. Phys. 24 (1953) 731.

15. Longuet-Higgins, H. C., and Roberts, M. deV., Proc. Roy. Soc. (London) 230A (1955) 110.

16. Talley, C. P., Post, B., and La Placa, S., This volume p. 83.

THE PREPARATION OF HIGH-PURITY BORON
BY HOT-WIRE TECHNIQUES

C. F. Powell*, C. J. Ish†, and J. M. Blocher, Jr.*

Techniques are described for preparing massive, 99.9+% pure boron in 100-g quantities by hydrogen reduction of boron tribromide vapor on tantalum filaments at 1450°C. Deposition rates up to 9 g/hr, with better than 90% recovery of boron from the tribromide, were obtained. Under optimum conditions, the impurity levels in parts per million in the product were as follows: oxygen 33, nitrogen 30, sulfur 50, silicon less than 300, carbon 200, tantalum (filament) 200, and other metals less than 100.

No appreciable interdiffusion was observed at the filament–deposit interface. In some cases, the filament was separated mechanically.

The hot-wire technique for preparing high-purity metals has undergone extensive development and application at Battelle Memorial Institute since 1943. Its use in the preparation of high-purity boron was studied in 1949 in connection with an evaluation made by the RAND Corporation to determine the potential of boron as a structural material. Further development of the process and techniques was made under the sponsorship of the Carborundum Company and the General Electric Company in connection with alloy development work.

The preparation of massive boron by hydrogen reduction of boron halides on a heated filament was selected after considering alternate methods of preparing high-purity boron. Boron prepared by fused salt electrolysis is in granular form and is contaminated with electrolyte, oxide, and possibly with high-melting nonvolatile borides. Reduction of the boron halides or oxide with calcium, magnesium, aluminum, or the alkali metals is unsatisfactory for the same reasons. Thermal decomposition of the boron halides in an electric arc can produce high-purity boron, but at a very low production rate, and in the form of fine powder which can become appreciably contaminated with oxygen. Hydrogen reduction of boron tribromide in a heated tube likewise can produce high-purity mate-

---

*Battelle Memorial Institute, Columbus, Ohio.
†Chemical Abstracts, Columbus, Ohio.

rial, but in easily contaminated powder form. Thermal decomposition of the boron hydrides on a hot filament might also be an effective way of preparing high-purity boron were it not for the extreme flammability and toxicity of these compounds, which renders impractical the production of appreciable amounts of boron.

Boron tribromide was selected for use in the hot-wire hydrogen reduction process because it is slightly easier to purify, especially of silicon- and carbon-bearing impurities, than the trichloride or trifluoride; it is easier to handle, being somewhat similar to water in physical properties; it can be reduced with hydrogen at slightly lower temperatures than the trichloride or trifluoride.

Reduction of the boron tribromide was carried out on tantalum filaments, since available data indicated that tantalum and tungsten showed the lowest diffusivity in boron of any of the potential filament materials, at temperatures below 1500°C. Since accurate data on the diffusivity of tantalum in boron were not available at the time this work was started, it was decided that the filament diameter should be kept as small as possible to minimize contamination of the deposit from core wire. Most of the deposits were obtained on 9- to 10-in. lengths of 3-mil-diam annealed tantalum wire. A few preparations were made on 1.1-mil-diam unannealed tantalum wire, but it was much more difficult to load this wire into the apparatus without kinking or breaking.

In the initial work in developing a satisfactory apparatus for carrying out the hot-wire reduction of boron halides, hydrogen reduction of both boron trichloride and tribromide was tried. Vertical filaments, either a single straight length or hairpins, were used. It was not possible, however, by this technique, to obtain deposits larger than about 3 to 5 mm in diameter. The top part of a deposit grown on a vertical wire invariably developed a spiny, discontinuous structure which resulted in overheating and eventually burnout at this part. This difficulty is believed to have been caused by excessive preheating of the deposition atmosphere as it was carried by convection currents over the length of the wire, or by the depletion of boron tribromide in this atmosphere. Spiny growth of the deposit on vertical filaments could be avoided by limiting the filament temperature to 800 to 900°C, but the deposition rate at these temperatures was extremely low, and the deposit was so brittle that small vibrations in the apparatus snapped the filament before it had grown to appreciable size.

Deposition on horizontal filaments was then tried as a means of limiting the path of convection currents over the wire. This technique was immediately successful, and permitted large boron deposits to be grown on very fine wires. The use of horizontal filaments com-

plicated the deposition apparatus somewhat, in that movable electrodes had to be provided to accommodate expansion and contraction
of the filament, and the filament had to be loaded with an adjustable
tension sufficient to prevent sagging as the filament grew, but in-
sufficient to cause breakage or excessive vibration.

Several versions of horizontal-filament deposition apparatus
were tried. The one most easily assembled, dismantled, and cleaned
is shown in Fig. 1. The essential parts of the apparatus were the
vaporizer, where hydrogen was bubbled through liquid boron tribromide; the deposition chamber; and the condenser, where unreacted
boron tribromide was removed from the waste gas and returned to
the vaporizer through a U-loop. The condenser was cooled to about
−20°C by alcohol which was circulated through a heat exchanger in
a separate dry-ice bath, or by cold brine from a mechanical refrig-
eration unit. All parts of the apparatus, except the deposition chamber, were constructed of Pyrex glass. The deposition was constructed
of Vycor. The ball joints and stopcocks were lubricated with Kel-F
fluorocarbon grease, which maintained its fluidity in contact with
boron tribromide vapor for several days before becoming gummy.
In another version of the apparatus, copper tubing connections with

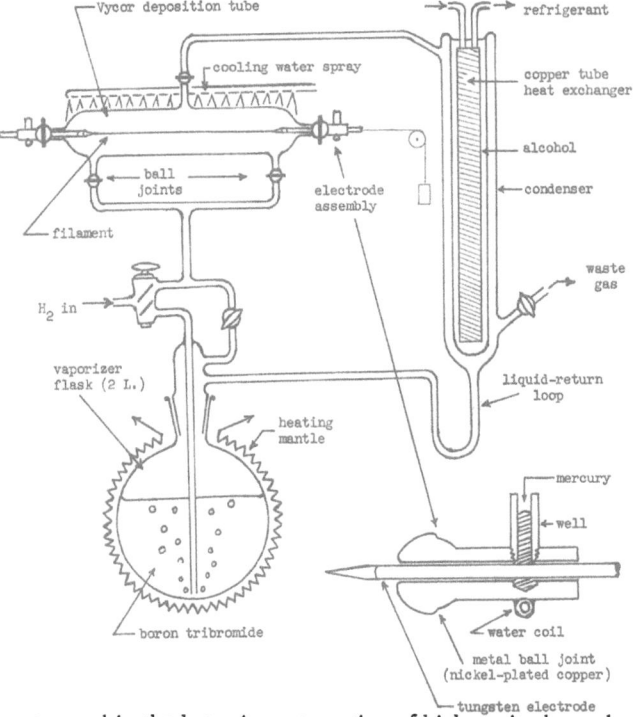

Fig. 1. Apparatus used in the hot-wire preparation of high-purity boron by hydrogen reduction of boron tribromide.

brass compression fittings and brass needle valves, sealed with Teflon gaskets, proved highly satisfactory.

Filaments were attached to 0.125- to 0.25-in.-diam tungsten rod electrodes through tapered sections cut from tantalum sheet. These tapered sections permitted a smooth transition of the boron growth from the fine filament to the tungsten electrode. The electrodes passed through water-cooled, nickel-plated copper ball joints clamped in the ends of the deposition tube. The sliding fit between the electrode and the ball joint was sealed with a mercury well, as shown in Fig. 1.

The 3-mil-diam tantalum filaments were heated from two power supplies. At the start, the filament was fed directly from a 110-v, 100-amp, variable transformer. After running until the filament voltage had dropped to about 35 v (from a starting value of about 80 v), and the current had risen to 80 or 100 amp, the filament leads were quickly switched to the output of an 8-kv-a, 200-amp transformer fed from a 220-v, 50-amp, variable transformer. This switching, by means of a large-capacity double-pole double-throw knife switch, was done so quickly that virtually no drop in filament temperature occurred.

The hydrogen used in the reduction was purified in two ways:
a. By passage of 99.8% electrolytic tank gas over copper turnings heated to 700°C, drying with magnesium perchlorate and calcium hydride, and a final passage over magnesium turnings heated to 600°C.
b. By passage of 99.9+% prepurified tank gas through a Deoxo catalytic combustion unit and drying with anhydrous calcium sulfate and a liquid nitrogen trap.

No variations in product purity attributable to the different methods of hydrogen purification were detected.

The boron tribromide used was obtained in two ways. In the initial work, crude amorphous boron, containing about 9% magnesium salts and other impurities, was brominated with dry, chemically pure bromine at 700°C in a fused silica tube. The product was purified by fractional distillation under an inert atmosphere in a 1-in.-diam. 3-ft-long, vacuum-jacketed Vigreaux column. A middle cut of the distillate, having a boiling point of 90°C, and amounting in volume to 70 to 80% of the initial volume of crude boron tribromide, was used in the reduction.

In the more recent work, "technical-grade" boron tribromide, supplied by the American Potash & Chemical Corporation, was used. The typical analysis given for this material was:
Boron tribromide. . . .99.5%
Free bromine 0.1%

Silicon as Si 0.001%
Sulfur as S_2Br_2. . . . . 0.04% max
Carbon as $COBr_2$ . . . . 0.15% max

This material was distilled in a 55-theoretical-plate Fenske helix-packed column having a capacity of 1.8 liters per hour. A center cut having a boiling point of 89 to 90°C, and amounting in volume to 60 to 67% of the initial volume of boron tribromide, was taken off at 40% reflux. This center cut was divided into three or four sections for use in boron preparation. It was noted that boron obtained from the first part of the center cut was higher in carbon contamination than the boron obtained from the latter portions of this cut.

The conditions used in depositing boron on a hot filament were as follows:

Filament temperature: 1300 to 1430°C
Boron tribromide vaporizer temperature: 50 to 70°C
Hydrogen flow rate: 5 to 9 ft^3/hr

The most critical period in the growth of boron deposits occurred in the first few minutes of deposition. Vibrations set up in the growing filament, probably from the ac heating current, easily snapped the small boron deposit unless rapid changes were made in the tension on the filament.

Control of filament temperature was by manual variation of the input voltage. Once the filament had become relatively pure boron (when the filament had increased from 3 to about 60 mils in diameter), constant regulation of the filament current was required to prevent the temperature from "running away" because of the large negative temperature coefficient of resistance of boron. As the deposit became larger, it showed less tendency to "run away" thermally, because of the large thermal mass present. However, loss of temperature control, especially if the filament temperature dropped to below 1200°C, usually caused large-diameter deposits to shatter from thermal shock. Automatic control of the filament temperature would have been highly desirable in these depositions.

Filament growth was limited to a maximum diameter of 0.5 to 0.75 in. by the power available for heating the filament, and by the capacity of the boron tribromide vaporizer. (A boron filament could not be shut down and then restarted as is done with most metallic filaments, since its cold resistance was of the order of megohms.) The maximum filament currents used were from 200 to 250 amp.

On a 9-in. filament the rate of boron deposition generally ranged from 8 to 10 g per hour. Plots of filament current versus time for several preparations indicated that the rate of boron deposition was nearly independent of the filament diameter throughout the deposition period. The deposition rate was determined chiefly by the filament temperature and the rate of flow of gas through the deposition tube, i.e., the deposition rate was determined primarily by the chemical equilibria existing at the filament. By condensing and recirculating the unreduced boron tribromide, over 90% of the boron contained in the tribromide was recovered in the deposit.

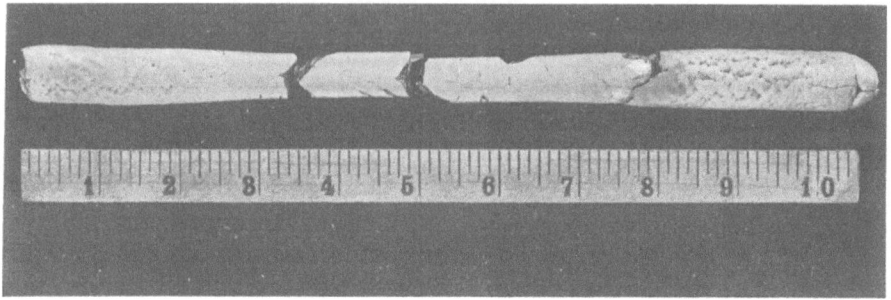

Fig. 2. Boron deposit No. 4506-23 (scale in inches).
Weight - 102.5 g   Purity - 99.9 + %

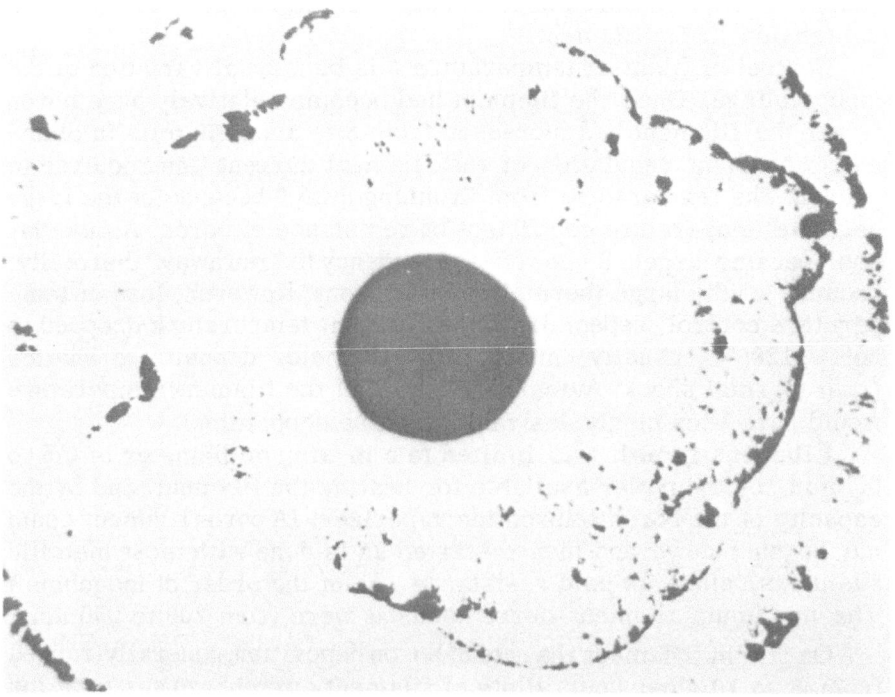

Fig. 3. Cross section of a hot-wire boron deposit (50×).

A picture of a typical deposit is shown in Fig. 2. A cross section of a similar deposit is shown in Fig. 3, in which it appears that some porosity does exist in the deposit. This occurred chiefly at the ends of the filament, near the vapor feed inlets, and appeared to be caused by the growth of club-shaped crystals or structures which eventually bridged over. These growths were distinctly different from the spiny growths which caused filament burnout in the earlier work. It is also apparent from Fig. 3 that little or no diffusion of the core wire into the deposit occurred. This would permit the amount of tantalum contamination of the deposit to be greatly reduced by splitting or crushing the deposit and leaching out the tantalum with a nitric—hydrofluoric acid mixture, with which massive boron is only slightly reactive.

The levels of impurities found in this boron are given in Table I. The analytical methods used were as follows:

Carbon: combustion, absorption of CO_2, conductimetric determination of the CO_2. Sensitivity, 0.00X ± 0.001%.

Sulfur: combustion, absorption of SO_2, titration with iodate. Sensitivity, 0.00X ± 0.001%.

Metallics and silicon: spectrographic analysis. The direct arc of the specimen in the crater of a spectrographic carbon rod was analyzed by a 1.5-m diffraction grating. Synthetic standards were made up by addition of impurities to boron from the best lot. Tantalum was calculated from the weight of the filament used.

Nitrogen: standard micro-Kjeldahl. Sensitivity, 0.00X ± 0.001%.

Oxygen and hydrogen: in the vacuum-fusion analyses, the sample, wrapped in degreased tin foil, was dropped into a carbon-saturated iron bath at 1200°C, and the temperature was quickly raised to 1650°C. The evolved gases were collected and analyzed by a low-pressure fractional freezing method. Sensitivity, ± 7 ppm oxygen and ±1 ppm hydrogen with a 1-g sample.

It is apparent from Table I that the chief impurity was carbon, the amount of which varied considerably from one preparation to another. The boron tribromide obtained from the two different sources previously mentioned each produced boron having about the same range of carbon contamination. The commercially supplied boron tribromide, before purification by fractional distillation, produced boron containing 0.8 to 1% carbon. With the purified boron tribromide, boron produced from the later cuts of a distillation contained less carbon than did that prepared from the earlier cuts. It was also noted that the carbon content of deposits obtained from one set of distillation cuts tended to vary inversely as the rate of take-off of the cut. This could result from the slow decomposition of a miscible, nonvolatile contaminant into volatile fragments.

Table I. Typical Analysis of Boron
Produced by the Hot-Wire Reduction
of Boron Tribromide

Impurity	Percent
Carbon	0.02 to 0.08
Sulfur	< 0.01
Silicon	< 0.03
Tantalum	0.02
Other metallics	< 0.01
Nitrogen	< 0.003
Oxygen	< 0.01
Hydrogen	0.02
Boron (by difference)	99.9+

The boron deposits had a light gray surface color and a glassy, black appearance at fractures, which tended to be conchoidal. One specimen showed a Knoop hardness at room temperature of 2160 ± 30 with a 500-g load. The melting point of the boron was about 2100°C (determined with an uncalibrated optical pyrometer and uncorrected for absorption by the window glass). The boron was extremely brittle at all temperatures up to at least 1200°C. Slight plasticity may have developed at higher temperatures as indicated by the fact that thick boron filaments developed a pronounced sag when operated for several hours at 1300 to 1400°C.

These boron deposits were not suitable for crystal structure work. The x-ray diffraction pattern showed about 40 lines, which did not match those reported by Laubengayer and co-workers [1] or the patterns of various crystal structure nets. This tends to corroborate the conclusions of others that two or more different crystal lattices can coexist in boron, and that the lattices are stable over a wide range of lattice imperfections.

Reference

1. Laubengayer, A. W., Hurd, D. T., Newkirk, A. E., and Hoard, J. L. J. Am. Chem. Soc. 65 (1943) 1924.

THE MANUFACTURE OF BORON

G. H. Fetterly[*]

In 1944 the Norton Company put into operation a small plant for the manufacture of massive crystalline boron. The capacity of this plant was approximately 300 lb per month on continuous operation. The product averaged 96-98% boron and contained 1-3% carbon with minor amounts of iron and silicon.

The central item of equipment comprised six 2-in.-diam graphite rods, 4 ft long, which were mounted vertically, connected in series, and heated to a temperature of 1400°C by a current of 4000 amp at 100-150 v. The rods were enclosed in a vertical cylindrical shell having a glass window for optical pyrometer readings. The shell was cooled by a continuous cascade of water. The coaxial well at the bottom of the shell was filled by polychlorpropane.

The furnace was fed with a mixture of boron trichloride (in gas form) and hydrogen. About 1 lb/hr of boron was deposited on the hot graphite rods on which the boron layer could sometimes be built up to a thickness of ½ in. When the furnace was cooled, the boron layer was broken away mechanically from an intermediate zone of high-boron boron carbide which formed by diffusion. The boron lumps had a shiny-black, conchoidal fracture. The density was 2.33 g/cm^3. The color in thin sections was yellow to red with a refractive index for lithium light of 2.5.

Boron trichloride was recovered from the effluent gases in a column packed with crushed dry ice.

The process is described in U.S. Patent No. 2,542,916 issued February 20, 1951, to G. H. Fetterley and assigned to the Norton Company.

Some of the difficulties of the operation and subsequent improvements are described. The importance of the purity of the boron trichloride to the quality of the product is noted and methods for the production of suitable boron trichloride are described.

The object of this paper is to outline the process used to make several hundred pounds of massive crystalline boron of 95% or

---

[*] Deceased; paper presented by G. R. Watson, Norton Company, Chippawa, Ontario, Canada.

higher purity for the use of the Manhattan District Corps of Engineers. This process hitherto produced only a few grams of boron under laboratory conditions. The semicommercial process that was developed, and which is described herein, was operated on a scale that produced 300 lb of boron per month.

Process Summary

The process is based on the reaction

$$2BCl_3 + 3H_2 \longrightarrow 2B + 6HCl.$$

This reaction has hitherto been carried out by passing the reactants over fine, electrically heated wires of a refractory metal such as tungsten, or through an arc between water-cooled electrodes. It has been made commercially useful by the development of a furnace in which the reaction is carried out on electrically heated graphite rods. The feasibility of this operation depends to a large extent on the mechanical arrangement of the rods to allow the use of reasonable values of current and voltage and to avoid failure due to thermal stresses.

As is indicated in the process summary chart (Fig. 1) a mixture of H_2 and BCl_3 is passed over a series of electrically heated graphite rods. Part of the BCl_3 reacts to form a dense, finely crystalline layer of boron which builds up slowly on the rods. At the conclusion of the furnace run, the boron is removed from the rods, separated from adherent graphite and boron carbide by sorting, and is then crushed, packaged, and shipped.

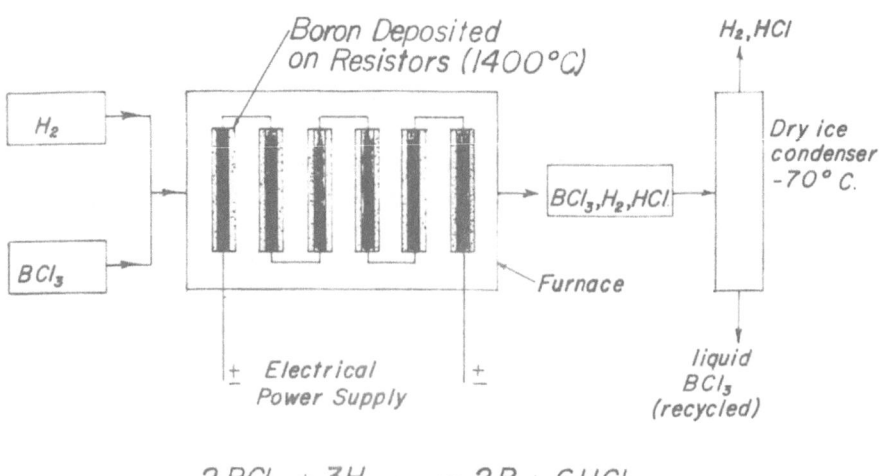

Fig. 1. Process summary.

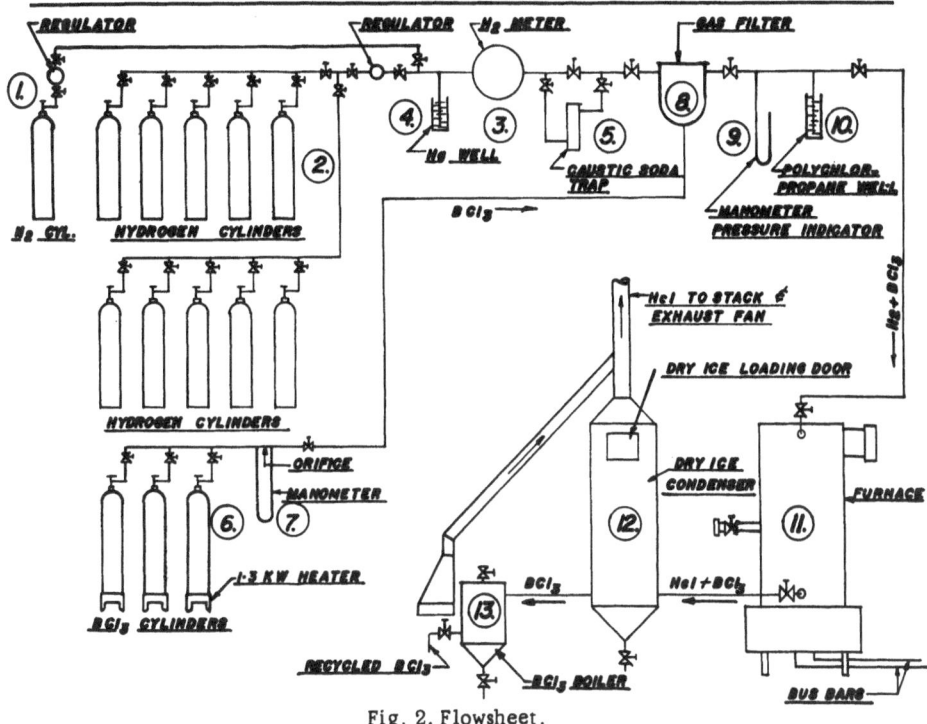

Fig. 2. Flowsheet.

The process is described in U.S. Patent No. 2,542,916, issued February 20, 1951, to G. H. Fetterley and assigned to the Norton Company.

Furnace Equipment and Operation

Figure 2 is a flowsheet for the furnace and associated equipment.

Before starting the run, nitrogen is taken from a nitrogen cylinder and regulator (1) to sweep air out of the system and prevent explosions. When running, hydrogen is taken from a manifold and regulator (2) through a 1-in. line to the integrating gas meter (3). A pressure relief valve (4) is connected to this line close to the meter inlet. A 1-in. line carries the hydrogen through a caustic soda trap (5) to a pipeline-type gas filter (8). BCl_3 is taken from a manifold (6), through a flowmeter (7), and mixes with the hydrogen in the gas filter (8). The mixed H_2 and BCl_3 pass through a 2-in. line to the top of the furnace (11). A manometer pressure indicator (9) and a pressure release valve (10) are connected between the filter and the furnace. A 2-in. line carries the effluent gases to the condenser (12), and waste gases pass up a 12-in. stack to the 900 ft^3/min ventilating fan. The condenser consists of a thermally insulated vertical shell with a cylindrical bottom. The shell is kept filled with crushed dry ice, supported on a screen of closely spaced steel bars welded across

Fig. 3.

Fig. 4.

the inside. The condensed BCl_3, with considerable HCl in solution, runs continuously through a 2-in. line into the boiler (13), where its temperature rises to that of the surroundings, and practically all the HCl is driven off. A sight glass on the boiler shows the liquid level. Liquid BCl_3 is drawn off from the boiler periodically into tanks packed in dry ice. These are later allowed to warm up to room temperature and put back on the BCl_3 manifold.

Figure 3 shows a photograph of the feed line as far as the 2-in. pipe leading to the furnace. This pipe is visible in the upper left-hand corner.

Figure 4 shows the furnace with part of the feed line on the right and the condenser on the left. Figure 5 shows the furnace set up for operation with the cooling water turned on.

Fig. 5.

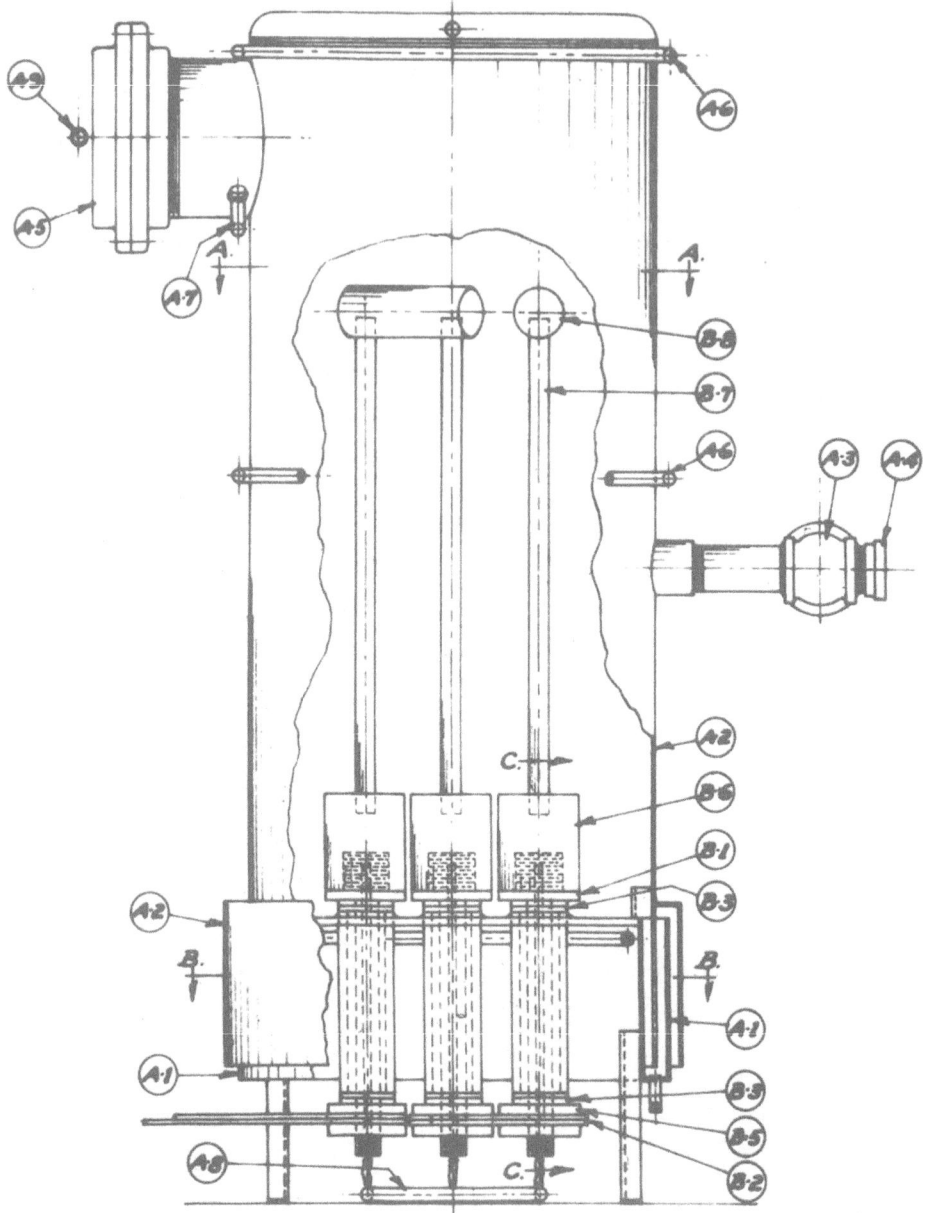

Fig. 6. Resistor furnace assembly.

Figure 6 shows an assembly drawing of the furnace, which is about 3 ft 4 in. in diameter × 8 ft 4 in. high. The furnace base (A-1) is of ¾-in. steel, surrounded by a well used as a liquid seal. The well has a cooling water spray on the outside. Six water-cooled bronze clamps pass through the base and are insulated from it. They terminate at their upper ends in cylindrical graphite blocks, B-6, into which the graphite rods, B-7, are fitted. Each electrode has a central hole, extending from the bottom almost to the top, and is cooled by a rapid flow of water from a coaxially mounted, ⅛-in. pipe extending almost to the end of the closed hole. The electrodes are connected at the top with graphite yokes, B-8. Bars and yokes are staggered so that all the rods are connected in series. The power bus bars are connected at any convenient point in the series circuit. The shell (A-2) is ¼-in. sheet steel, has a capacity of 100 ft^3, and weighs 1500 lb. It is fitted with:

a. a skirt to keep water out of the well (A-2, left side of drawing)
b. a sight tube for temperature measurement (A-4)
c. a 16-in. lead-coated safety diaphragm (A-5)
d. a strong eye at the top for the hoisting hook (not shown)
e. two water-spray rings around the shell at the top and center (A-6)
f. water sprays on the underside of the diaphragm offset and and on the diaphragm (A-7 and A-9, respectively)
g. 2-in. inlet and outlet pipe connections (not shown)

Operating Conditions Summary

Operating conditions were a compromise between rapid production and high yield. The nature of the reaction makes these objectives to some extent mutually exclusive. The BCl$_3$ was set at the highest rate that would allow an over-all gross yield of crude of 70%. The H$_2$/BCl$_3$ ratio was as high as the hydrogen handling facilities would permit. The condenser efficiency was affected by the weather, due to condensation in the stack and on the dry ice. Conditions of operation are summarized below:

Rod temperature . . . . 1400-1450°C
H$_2$ feed . . . . 600 ft^3/hr
BCl$_3$ feed . . . . 40 lb/hr
Mol ratio H$_2$/BCl$_3$ 5
One-pass yield of crude 27%
Gross crude production rate . . . . 1.0-1.1 lb/hr
Condenser efficiency 83-87%
Over-all yield of crude. 70%
Over-all yield of finished product. 40%

Figure 7 shows the interior of the furnace at the conclusion of the run. Note the layer of boron on each rod.

The length of run was highly variable, largely because of the tendency of the rods to break at the ends. Anything under 20 hr was considered very poor, and 60 hr or longer was unusually good. Generally, however, the boron would build up as a dense layer to a

Fig. 7.

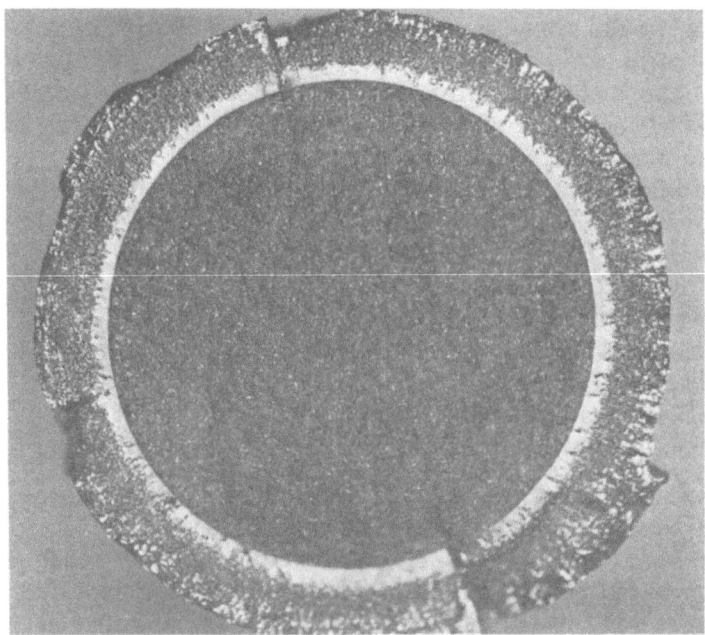

Fig. 8.

thickness of ½ in., at which point one or more of the rods would break from thermal stresses. The large difference between the yields of crude and finished product was due in part to the severe conditions imposed on the sorters to obtain a product having a purity exceeding 97% B. If 95% purity were sufficient, the net yield could be increased considerably. Figure 8 is a cross section of a graphite rod which clearly shows the carbide layer adjacent to the graphite and the outer layer of boron.

Operating Difficulties

PLUGGING OF GAS LINES. BCl_3 is such an active drying agent that it was found to be practically impossible to dry the hydrogen thoroughly enough to prevent deposition of boric acid at the point where the H_2 and BCl_3 mix. It was found that a mixing chamber at this point caught practically all the boric acid.

Fog in the furnace indicated the presence of oxygen or water. It was usually followed by plugging of the line from the condenser. The latter line plugs occasionally in normal operation, usually at the end inside the condenser.

A closely related difficulty was fogging of the glass of the sight tube. This occurred almost imperceptibly and could cause a large error in temperature reading. The sight tube was sealed off at frequent intervals with the gate valve provided, and the glass cleaned.

SLUDGE FORMATION. A thick sludge of boric acid and dilute HCl tended to collect in the bottom of the fan casing, which was remounted so that the sludge could run out by gravity. Condensate from the stack and hoar frost from the dry ice tended to form a plug in the center of the condenser. This was controlled by breaking up any incipient plugs in the dry ice with a long bar. Some sludge always formed in the condenser and settled to the sump at the bottom. This was cleaned out periodically whenever suspended solids began to appear in the recycled BCl_3.

EXPLOSIONS. Side reactions produced small quantities of spontaneously flammable or explosive compounds, which made it extremely hazardous to have hydrogen and air in the furnace simultaneously. Several explosions occurred before a standardized starting procedure was adopted.

CORROSION. The rapid corrosion of equipment was a serious problem. The welds in the sheet metal parts, especially the exhaust stack, were attacked rapidly. The fan should be made entirely of corrosion-resistant material.

ROD FAILURE. The most persistent difficulty was premature breakage of the graphite rods near the ends, probably due to thermal stresses within the rods. No correlation was found between "the tendency to break" and the bending strength. This seems to be a pinching-off effect as described by P. N. Bridgman [1]. Designs for slotted rods which gave promising results were evolved. It is believed a further study of the effect of variation in slot distribution will solve the problem.

Crude Processing Equipment and Operation

FLOWSHEET. Since the flowsheet is very simple, a drawing is not included. The product was separated from graphite and boron carbide by hand sorting, crushed to pass a 60-mesh screen, passed through a magnetic separator, and packaged for shipment.

The yield of final product was a function of the purity required and was consistent at 60% for an average purity of 97%. The yield of this process can be raised to 75-80% if the purity is lowered to 95%.

OPERATING DIFFICULTIES IN PROCESSING CRUDE. In the furnacing process, if the temperature is held above 1400°C and the full feed is maintained without interruption throughout the run, the layers of boron and boron carbide already described will be well defined and easily separated. If the temperature is too low, the layers are poorly defined and very adherent. Feed interruptions produced a layered structure, in which the boron itself broke into small pieces and flakes, and greatly increased the labor of sorting.

Effect of Purity of Raw Materials on Quality of Product

HYDROGEN. At maximum production rate, hydrogen consumption was 600 ft^3/hr, or averaged 12,000 ft^3 per day with 15% shutdown time. Little is known about the effect of impurities in the hydrogen. Analysis of one cylinder used on this project showed 0.2% oxygen, which was considered satisfactory.

BORON TRICHLORIDE. BCl_3 is a colorless liquid of density 1.4 g/cc, boiling at about 13°C at atmospheric pressure. The vapor fumes strongly in moist air and is irritating, but not appreciably toxic. The liquid was received in standard chlorine cylinders, holding 100 or 150 lb. At maximum production rate, net consumption was 17 lb/hr. Average requirements were 350 lb per day.

In subsequent years, we actually manufactured our own high-purity BCl_3 by chlorinating boron carbide in furnaces of our own design.

The most important impurities in BCl_3 are $SiCl_4$, Cl_2, and $COCl_2$. $SiCl_4$ forms silicon in the furnace, the percentage of Si in the product being about twice the percentage of $SiCl_4$ in the BCl_3. An upper limit of 0.2% was found practical for this impurity. Chlorine is quite soluble in BCl_3, and its presence causes an explosion hazard. About 0.02% is sufficient to give the otherwise colorless liquid a yellow color. All the BCl_3 used on this project was substantially free of chlorine, and it is not known how high the chlorine content must be to constitute a hazard. Phosgene, $COCl_2$, is an undesirable contaminant of BCl_3, because a high proportion of its carbon content is co-deposited with the boron and yields a high carbon product. BCl_3 containing 2.5% phosgene yielded products with 3-4% carbon. An upper limit of 0.2% phosgene was considered mandatory, and BCl_3 with lower phosgene content gave products of good quality (95-99% B, 0.3-2% C).

Major Supplies

GRAPHITE RODS. Six 2 in.×48 in. graphite rods were required for each run. Normal consumption was 24 rods per week.

DRY ICE. Consumption of solid carbon dioxide was about 1500 lb per day. It was found convenient to have this delivered once a day in 50-lb cakes. Shortly after delivery, it was crushed in a standard jaw crusher to about 1-in. pieces and stored in an insulated box.

ELECTRICAL POWER. The power supply was capable of delivering 4000 amp, with continuous control of voltage between 100 and 175 v. Power demand did not normally exceed 600 kv-a, but conservative design used equipment rated at 750 kv-a.

Products

BORON. The principal product was boron of 95-99% purity, in long, thin strips, and in black, finely crystalline lumps with a shining conchoidal fracture. These were normally mixed together and crushed to pass a 60-mesh screen. A typical analysis of the product is shown below:

B.	97.65%
C.	1.29
Fe.	0.22
Al	0.03
Ca.	0.27
Mg.	<0.05
Cu.	trace
Mn.	"
Si	"

The x-ray powder pattern is similar to that published by Laubengayer et al. [2] for needle crystals of boron. Crystals up to 35 μ were observed under polarized light. The color, in thin sections, is yellow to red. The refractive index for Li light is about 2.5. The density, for 98-99% purity, is 2.33 g/cm^3.

BORON CARBIDE. The principal by-product is high-boron carbide, containing 80-90% boron. It occurs between the high-purity boron and the graphite rod as a dull black cylindrical section with a sandy structure, and breaks off in curved plates which may be as long as 2 in.

TAILINGS. When the boron and boron carbide have been separated, there always remains a finely divided mixture of boron, boron carbide, and some graphite. Much of the boron can be identified by its bright, fractured surfaces and picked out with tweezers, but the process is excessively laborious.

Outline of Recent Improvements

In recent years, interest in boron has revived, and as a result, the Norton Company has redesigned the original boron process to improve yields and reduce costs.

New designs of chlorination furnaces were tested for the manufacture of BCl_3 from Cl_2 and B_4C. These proved to be simple and practical.

In the original process, the unreacted hydrogen was allowed to go to waste. In the improved plant, this raw material is recirculated after removal of HCl and BCl_3. New designs of condensers are now used to improve yields and lessen downtime. The same principles are in force, however, and the process is essentially the same as described in this paper.

Summary

This paper has briefly outlined a practical process for manufacturing elemental crystalline boron which was developed at Norton Company by the author. The process is of historical interest at least, because it was the first time that crystalline boron had been made in large quantities. Although some of the equipment was admittedly crude, it was nevertheless effective, and the extremely urgent need for the product by the U.S. Army was a spur to the whole project. Refinements have been made, but the general process has not been changed in principle.

References

1. Bridgman, P. N., J. Appl. Phys. 9 (1938) 517.
2. Laubengayer, A. W., Hurd, D. T., Newkirk, A. E., and Hoard, J. L., J. Am. Chem. Soc. 65 (1943) 1924.

PREPARATION OF BORON FROM BORON CARBIDE*

David R. Stern[†]

A process for the preparation of elemental boron from boron carbide is presented. Included is evidence which shows that the process is one of anodic transfer. Boron of at least 99.8% purity can be made directly from technical-grade materials. Purity is a function of electrolysis voltages and boron carbide purity. Current efficiencies in excess of 95% are obtained and preferred electrolysis conditions are stated.

Preparation of elemental boron has been discussed by a large number of investigators [1-19]. Not all processes give boron of sufficient purity. Filament and arc discharge techniques have generally been regarded as the preferred methods for securing a high-purity crystalline boron [2,3,5,6,7,8,11,12,17,19]. New interest in elemental boron as thermistors and possible semiconductor devices has reemphasized the problem of attaining higher purities. McCarty and co-workers [18] have discovered new low-temperature crystalline modifications of elemental boron.

Vapor phase deposition techniques are generally limited by metallic filament contamination. Impurities are generally tantalum, tungsten, molybdenum, carbon, and titanium [2,3,7,11,12,17,19]. Another disadvantage with these contaminants is that they are not removed with secondary upgrading techniques. Deposition on a filament of elemental boron and lower decomposition temperatures are significant contributions [7,19].

Desire for a product of potentially higher purity without heavy metal contamination focused attention on an electrolytic approach. Andrieux, Cooper, and others have developed electrolytic reductions [4,9,10,14,15,16].

This research effort has resulted in an electrolytic process for the preparation of boron which is an anode transfer technique [13]. This transfer technique has been applied both to elemental boron and boron carbide. Feasibility studies were conducted in the laboratory on an exploratory basis in small cells. The scope of this

---

* The editors wish to thank the Journal of the Electrochemical Society for permission to republish this paper.
† American Potash & Chemical Corp., Whittier, Cal.

27

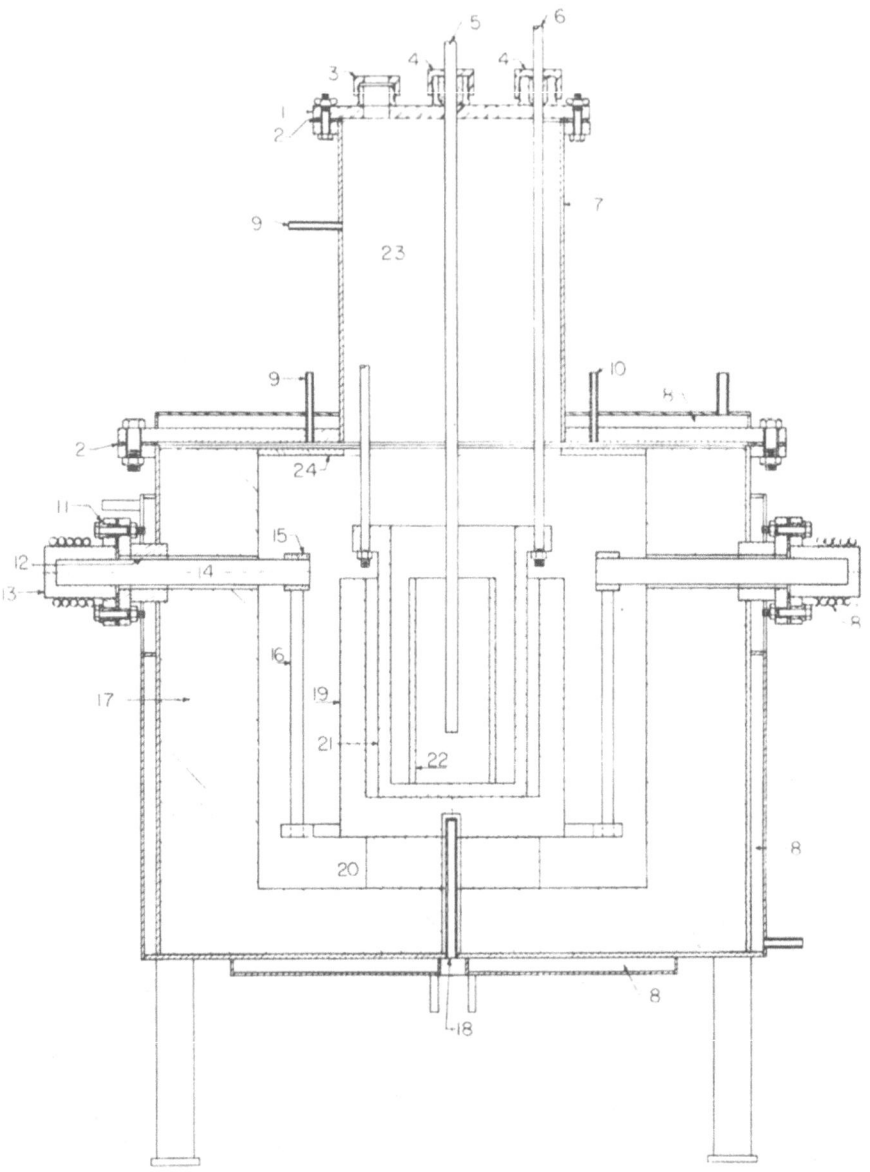

Fig. 1. Bench-scale boron carbide cell: 1) stainless (316) lid; 2) Neoprene gasket; 3) sight glass; 4) packing glands; 5) cathode; 6) anode tie rods; 7) mild steel shell; 8) water jackets; 9) argon inlet; 10) argon outlet; 11) insulating bushing; 12) Vycor tube; 13) copper ac electrode connector assembly; 14) main graphite electrode lead; 15) graphite ac electrode connector; 16) ac graphite resistor rod; 17) magnesia insulating bricks; 18) centering post; 19) dense carbon cup; 20) carbon pedestal; 21) porous carbon basket; 22) porous carbon diaphragm; 23) cooling chamber; 24) heat baffle.

paper will be limited to the bench-scale preparation of elemental boron from boron carbide.

Experimental Procedure

CELL. A cross section of the cell used in this investigation is presented in Fig. 1. The unit consisted of a porous carbon basket with a porous carbon diaphragm. Carbide was packed between the diaphragm and porous basket. This cartridge unit was lowered into a dense graphite crucible by steel rods. Removal and recharging with boron carbide could be accomplished without a complete cell shutdown.

External heating was accomplished using graphite resistors. Cell temperatures were measured with chromel-alumel thermocouples inserted into the recesses in the centering post and the hollow cathode which was also a thermowell.

An argon blanket could be maintained over the entire cell.

ELECTROLYTE. The electrolyte consisted of sodium chloride, potassium chloride, and potassium fluoborate with an initial composition of 40-40-20 wt.%, respectively. During the course of electrolysis, cathode deposits were periodically withdrawn. Occluded electrolyte was water leached and could be recovered and recycled.

ELECTROLYSIS. After the anode compartment was packed with boron carbide, sufficient predried salt was added to cover the carbide. This cartridge charge was then placed into the heating furnace and melted under argon. Whenever fresh salts, a new diaphragm, or a new crucible were used, pre-electrolysis was performed at 2.5 v from ½ to 1 hr before actual electrolysis. The pre-electrolysis deposit acted as a scavenger of impurities and was a direct means of purifying the electrolytic system. A cathode was then inserted to a known depth so that a given cathodic current density could be set initially.

Fig. 2. Typical cathode deposit (scale in inches).

At completion of electrolysis the cathode was raised and allowed to cool. To recharge the boron carbide the cartridge unit was raised, the salt allowed to drain, and a new unit inserted.

PROCESSING. Figure 2 is a photograph of a typical cathode deposit. On removal from the cell this deposit was placed in a polyethylene beaker of distilled water. Upon standing in the water the deposit would break off the rod and was then wet-ground in an iron mortar and leached twice with 10% hydrochloric acid. It was then washed free of acid, rinsed with acetone, and vacuum-dried under argon.

The final sample was then weighed and analyzed. Most samples were analyzed for total boron content, sodium, iron, potassium, silicon, nitric acid insolubles, and water-soluble boron.

Experimental Results

PRODUCT PURITY. The purest elemental boron obtained in this investigation was 99.8%. This purity was achieved using commercial reagents without any prior purification or recrystallization. Control of the electrolysis and atmosphere were the only precautions taken. Major impurities were sodium, silicon, iron, and carbon. These generally could be limited to less than a total of 0.2%. Lower levels

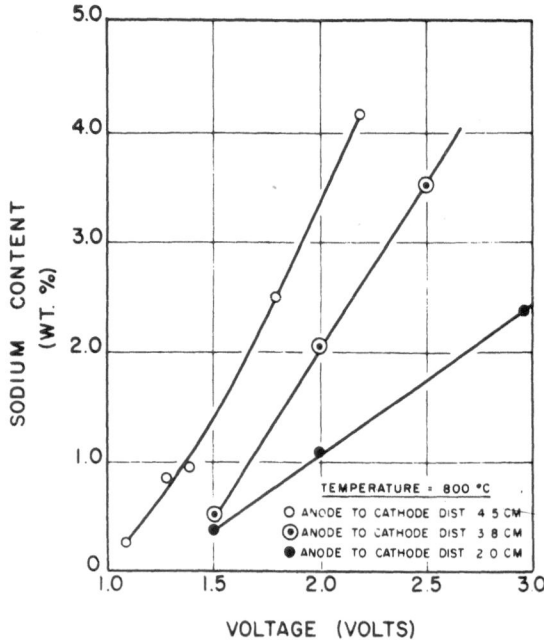

Fig. 3. Sodium content of boron product as a function of cell voltage.

of impurities were sodium 0.05%, silicon 0.07%, iron 0.05%, and carbon (as B_4C) 0.06%. These lower levels, however, were not obtained simultaneously in one sample.

EFFECT OF CELL VOLTAGE. An important factor contributing to product purity is cell voltage. A typical relationship between sodium impurity and cell voltage is presented in Fig. 3. At voltages between 0.9-1.0 v sodium contents were as low as 0.05%, but no smooth correlation was obtained. It must be pointed out that anodic current density is not constant in this system.

A similar relationship can be shown for the silicon and iron impurities, but the final levels of these impurities are also dependent on the processing technique.

These results indicate that the concentration of major impurities in the product is a direct function of cell voltage, i.e., the higher the cell voltages, the higher the impurities.

Electrolytic decomposition of the fluoborate and alkali halides occurs at higher voltages. These reactions are evidenced by the liberation of chlorine and adversely affect the current efficiency of the process as well as product purity.

As long as sodium is a component of the electrolyte, there will be some sodium contamination, but by maintaining the voltage below 1.2 v, sodium content may be kept below 0.2%.

ORIGIN OF OTHER CONTAMINANTS. Table I lists the typical spectrographic analyses of the new raw materials entering the process and of the materials of construction. Boron carbide contains the major contaminants except for the calcium in the fluoborate. Thus,

Table I. Semiquantitative Spectrographic Analyses (wt. %)

Constituent	B_4C *	KBF_4 *	Carbon †	Graphite ‡
Boron	major	major	-	0.0008
Silicon	0.17	0.02	0.017	0.012
Iron	1.2	0.015	0.045	0.032
Magnesium	0.26	0.00092	0.0068	0.0016
Titanium	0.10	nil	0.0018	0.0036
Copper	0.0072	0.0011	0.0013	0.00044
Manganese	0.16	nil	-	-
Calcium	0.026	0.26	0.038	0.018
Zirconium	0.20	nil	-	-
Aluminum	0.3	0.0034	0.012	0.030

*Technical grades
†Grade 20 porous carbon
‡C.S. graphite

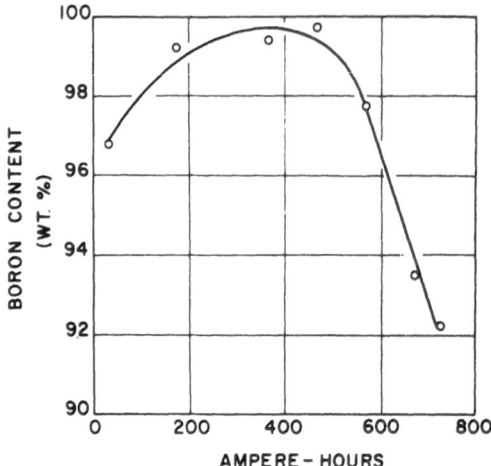

Fig. 4. Product purity as a function of cumulative ampere-hours.

purity of the elemental boron is closely allied to the purity of boron carbide. Silicon and iron are anodically transferred in this process.

Pre-electrolysis yields a deposit containing approximately 14% iron, 12% silicon, and only 16% boron.

Figure 4 presents product purity as a function of cumulative electrolysis. These results indicate that whenever a new charge is electrolyzed, the purity is low, probably due to moisture and impurities in the boron carbide and salts.

After an initial conditioning period the purity increases and remains high during most of the remaining electrolysis. Drop-off in purity is attributed to higher transfer voltages resulting from the buildup of carbon in the anode compartment as boron is depleted. If the product from 90% of the electrolysis time were combined into a single sample, an average product purity of 98% would result. The maximum purity attained was 99.8% boron.

EFFECT OF TEMPERATURE. A high-purity product is obtained at good current efficiencies within a temperature range of 700-850°C. No definite relationship between temperature and product purity has been noticed within this range. Although a detailed study of the phase relationship of the $KCl-NaCl-KBF_4$ mixture has not been made, operation above the $KCl-NaCl$ eutectic 660°C must be employed. Above 850°C, loss of the fluoborate increases and deposits tend to fall from the cathode. A compromise temperature of 800°C was adopted for the major part of this investigation.

CURRENT DENSITY. Cathodic current density has been varied between 12.0 and 75.0 amp/dm². High current efficiencies are ob-

Table II. Electrolytic Systems

Electrolyte	Anode	Voltage	Yield
KCl-KBF$_4$-NaCl	None	< 1.0	None
KCl-KBF$_4$-NaCl	None	>1.3	Some
KCl-KBF$_4$	None	< 1.0	None
KCl-NaCl	Boron carbide	2.2	None
KCl-NaCl	Boron carbide	3.2	None *
KCl-NaCl-NaF	Boron carbide	1.2	None
KCl-KBF$_4$	Boron carbide	< 1.0	Yes
KCl-KBF$_4$-NaCl	Boron carbide	< 1.0	Yes
LiCl-KF-KBF$_4$	Boron carbide	2.8	Some
KCl-KF-KBF$_4$	Boron carbide	< 1.1	Yes

* BCl$_3$ fumes

tained throughout this range. No significant correlation was found between initial current density and boron purity.

BORON MATERIAL BALANCE. It has been stated that boron was derived from the carbide and that fluoborate was only a carrier salt. Previous electrolysis systems [4,9,10,13,14,15,16] have operated at higher voltages and consequently have not been anodic transfer processes.

Table II lists experiments which offer some proof that in this system boron is derived from the anodic boron carbide.

The following points should be noted:

1. Experiments conducted without boron carbide (regardless of a carrier) yielded no product below 1.3 v.
2. Experiments conducted without fluoborate (with B$_4$C) also yielded no product, even up to 2.2 v.
3. The solvent for the carrier salt may be pure KCl, NaCl, or halide mixtures.
4. At voltages higher than 3 v boron trichloride is formed.

Analysis of the boron carbide also gives evidence of the boron being extracted. These data are presented in Table III.

Final proof of boron transfer is presented in Table IV. Examination indicates that a total of 61 g of elemental boron was extracted on the cathode deposit during electrolysis, whereas only 12 g of boron was available initially in the fluoborate and 11 g of this was accounted for at the completion of electrolysis.

CELL ELECTROLYTE. Cell electrolyte analyses have indicated that concentrations of sodium, potassium, and chloride ions do not change appreciably during electrolysis. Concentrations as a

Table III. Anodic Transfer of Boron from Boron Carbide
(in wt. %)

	Before electrolysis*	After electrolysis †
Total boron	75.4	18.8
Water-soluble boron	0.05	-
Nitric acid insolubles	98.2	83.58
Iron	0.11	0.04
Silicon	0.16	-
Boron in nitric acid insolubles	76.58	19.61

*Technical-grade powder B_4C

†Analysis after 1083 amp-hr of electrolysis

Table IV. Boron Balance in Boron Carbide Cell

Total boron in cell initially:
    As B_4C 74 g
    As KBF_4 (electrolyte)   12 g
                                              86 g

Total boron remaining in cell after
electrolysis:
    In anode charge   13 g
    In electrolyte   8 g
                                21 g

Total boron removed from cell during
electrolysis:
    By deposition   61 g
    By salt drag-out and sampling   3 g
                             64 g

Total boron accounted for   85 g
Boron unaccounted for   1 g

Table V.   Electrolysis Data

Temp. (°C)	Voltage (v)	Current (amp)	Time (hr)	Final deposit weight (g)	Cathodic current density (amp/dm²)	Current efficiency (%)	(kw-hr/lb boron)
735	1.1	6.9	24	21	16.6	92.0	3.9
740	1.1	6.5	23	18.8	14.4	90	4.0
725	1.1	3.5	29.5	12.0	11.9	81	4.3
790	1.0	7.9	19.0	17.4	13.5	96	3.9
800	1.0	6.9	19.0	17.5	14.0	98	3.4
800	1.0	4.2	24.0	12.7	9.0	90	3.6
795	1.0	5.3	23	17.1	11.3	97	3.2
800	1.0	3.7	23	12.9	8.4	101	3.0
810	0.9	5.8	52.5	39.6	10.7	90	3.1
810	1.0	4.1	66	35.7	8.2	94	3.4
800	1.0	5.7	30	21.7	5.1	92	3.6

function of cumulative ampere-hours are presented in Fig. 5. All concentrations are relatively constant except that of the fluoride ion. This is contrary to what is expected in the electrolysis of a halide—fluoborate system where the potassium fluoride concentration increases as boron fluoride is reduced to boron. At the start the boron concentration increases for a short period of time. This probably represents the solution of boron from the carbide. This observation is vividly illustrated in Fig. 6 when the fluoride to boron mole ratios are plotted as a function of electrolysis time. The ratio in fluoborate is four, while an inspection of the curve shows a decrease to a ratio of two. Chemical analysis of the purge gas indicates that some fluoride is lost from the system, possibly as boron trifluoride, but the rate of solution of boron from this carbide more than compensates for that volatilized from the system. Material balance data indicate that the physical quantity of fluoborate lost is relatively small and can be made up by fluoborate additions.

ELECTROLYSIS DATA. Current efficiency for this system is based on the following electrode reaction:

$$B^{3+} + 3e \longrightarrow B^0$$

Some data on electrolysis are presented in Table V. High current efficiencies are obtained and energy consumption is low. A pound of 94-99% boron requires from 3-4 kw-hr of electrical energy.

Summary and Conclusions

An electrolytic anode transfer process for the preparation of high-purity elemental boron from boron carbide has been presented.

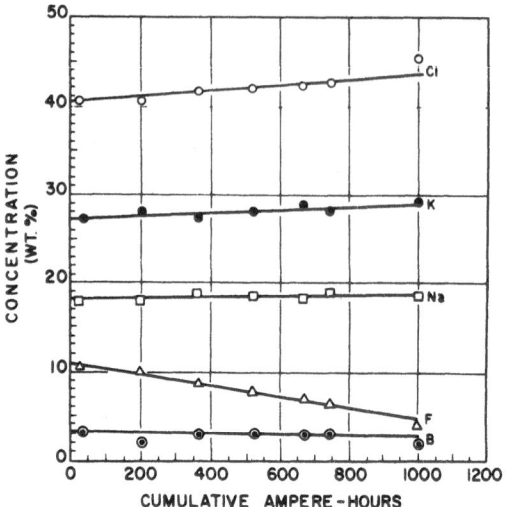

Fig. 5. Electrolyte composition as a function of cumulative ampere-hours.

Evidence that the process is one of anodic transfer is included.

The following conclusions concerning this method of preparing elemental boron can be set forth:

1. Boron of at least 99.8% purity can be made directly from technical grade materials without prior purification.
2. Elemental boron can be prepared at least equal in purity to that obtained by vapor deposition processes without the corresponding heavy metal contamination.
3. Major impurities are sodium, silicon, iron, and carbon. The total of these impurities can be reduced to 0.2% or less.
4. Impurity levels in the boron are a function of the electrolysis voltage and of the purity of the boron carbide.
5. Voltages less than 1.2 v, cathodic current densities between 12.0 and 75.0 amp/dm^2, and 800°C are the preferred conditions of electrolysis.
6. Current efficiencies for this anodic transfer process are high. Efficiencies in excess of 95% are easily obtained. Energy consumption is 3-4 kw-hr/lb boron.
7. The application of this technique as a secondary upgrading technique for boron of higher purity has been postulated.

Acknowledgment

The author would like to express his appreciation to Mr. Aiji Uchiyama and Mr. Quentin H. McKenna, who conducted both the laboratory and bench-scale work on this project.

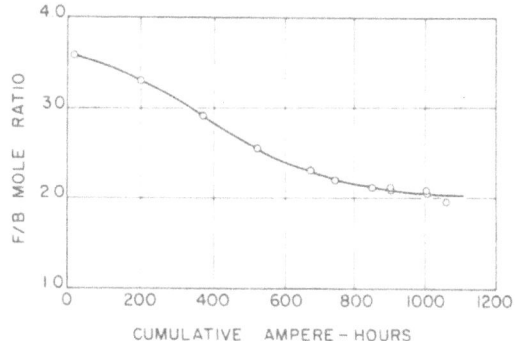

Fig. 6. Fluoride/boron mole ratio as a function of cumulative ampere-hours.

References

1. Moissan, H., Ann. chim. et. phys. 6 (1895) 296.
2. Weintraub, E., Trans. Am. Electrochem. Soc. 16 (1909) 165.
3. Van Arkel, A. E., U.S. Pat. 1,774,410, Aug. 26, 1930.
4. Andrieux, J. L., Ann. chim. 12 (1929) 423.
5. Hackspill, L., Stieber, A., and Hocart, R., Compt. rend. 193 (1931) 776.
6. Mellor, D. P., Cohen, S. B., and Underwood, E. B., Australian Chem. Inst. J. & Proc. 3 (1936) 329.
7. Laubengayer, A. J., Hurd, D. T., Newkirk, A. E., and Hoard, J.L., J. Am. Chem. Soc. 65 (1943) 1924.
8. Formstecher, M. and Ryskevic, E., Compt. rend. 221 (1945)747.
9. Cooper, H. S., U.S. Pat. 2,572,248, Oct. 23, 1951.
10. Cooper, H. S., U.S. Pat. 2,572,249, Oct. 23, 1951.
11. Fetterley, G. H., U.S. Pat. 2,542,916, Feb. 20, 1951.
12. Murphy, G. M., "Separation of boron isotopes", National Nuclear Energy Series, Vol. 3, U.S. Atomic Energy Commission, Oak Ridge, Tennessee (1952).
13. Norton Co., Australian Pat. 164,170, July 18, 1955.
14. Andrieux, J. L. and Deiss, W. J., Bull. soc. chim., France (1955) 836.
15. Nies, N. P., Fajans, E. W., Thomas, L. L., Hiebert, L. E., and Morgan, V., U.S. Pat. 2,832,730, Apr. 29, 1958.
16. Murphy, N.F. and Tinsley, R. S., U.S. Pat. 2,848,396, Aug. 19, 1958.
17. Stern, D. R. and Lynds, L., J. Electrochem. Soc. 105, (11) (1958) 676.
18. McCarty, L. V., Kasper, J. S., Horn, F. H., Decker, B. F., and Newkirk, A. E., J. Am. Chem. Soc. 80 (1958) 2592.
19. Bean, K. E. and Sparks, R. J., "Research investigation of physical chemistry and metallurgy of semiconductor materials", 2nd Quarterly Report, Mar. 1, 1959-June 1, 1959, Signal Corp, Contract DA-36-039-SC-78246.

PREPARATION OF CRYSTALLINE BORON

J. Yannacakis and N. P. Nies[*]

The method of preparation of crystalline boron by the U.S. Borax & Chemical Corporation involves (1) preparation of Moissan boron of about 89% purity, (2) upgrading the Moissan boron to 95% B by heating with fluorides, and (3) heating the upgraded boron by radiation in a vacuum. The latter step is accomplished by cold-pressing the powder into rings, stacking the rings to form a hollow cylinder, and heating in excess of 2000°C by means of a resistance heater in the center void. After cooling, the unfused material at the outside of the cylinder is removed mechanically. The fused portion is crushed to the desired particle size, screened, leached with HCl to remove iron, and then washed and dried. The product contains about 99.0% B.

The method of preparation of crystalline boron used by the U.S. Borax & Chemical Corporation involves three steps: (1) preparation of a suitable Moissan boron, (2) upgrading this product by heating with fluorides and leaching with acid, and (3) vacuum heating.

In the first step, magnesium powder or turnings is mixed preferably with about two parts powdered B_2O_3, reacted, and leached with acid to give a product containing typically 89% B, 8% Mg, and 3% other impurities.

In the second step the Moissan boron is upgraded to 95% B or better by fluoride treatment. The reaction vessel is an elongated steel cylinder with a removable cover at its upper end. It is provided with a sealing gasket and a cooling jacket for keeping the gasket cool while the lower part of the cylinder is heated. A mixture of the impure boron and the fluoride reagents is loaded into a container which is placed into the cylinder. The vessel is evacuated and flushed several times with argon or helium. The vessel is then heated to 950-1000°C for at least half an hour and cooled. The product is washed with hot acid, followed by water, and then dried; it contains 95-97% B. It is a powder and is generally lighter brown in color than the Moissan boron. Table I shows the effect of various reagent compositions on the product obtained. This table indicates that the most effective reagents for upgrading Moissan boron are KHF_2 and KBF_4. Less effective are KF, HF, and BF_3. The following reactions may

---

[*] U.S. Borax Research Corp., Anaheim, Cal.

Table I. Upgrading of Moissan Boron at 1000°C in Argon, 40 min

(89.1% B, 7.8% Mg, 3.1% impurities other than Mg)
(U.S. 2,893,842)

Run	Reagent composition (wt. % of Moissan boron)					Product		
	KBF$_4$	KHF$_2$	KF	BF$_3$	HF	% B	% Mg	% other
1	50					95.5	2.9	1.6
2		50				96.3	2.3	1.4
3	100					96.3	1.6	2.1
4		100				96.5	1.4	2.1
5	50	50				96.7	0.9	2.4
6	100	100				97.0	0.7	2.3
7	200	200				93.9	0.6	5.5
8			50			94.0	5.2	0.8
9			100			94.3	4.6	1.1
10	50	50	50			97.0	1.3	1.7
11				gas		94.1	3.5	2.4
12					gas	91.4	6.2	2.4

represent the upgrading process. Known standard free energies of formation are included.

$$2\,KHF_2 + MgB_{25}O_{0.5} = MgF_2 + 2\,KF + H_2 + B_{25}O_{0.5}$$

$\Delta F°_{298}$ −407.46   −250.8   −254.84

$$2\,BF_3 + 3\,MgB_{25}O_{0.5} = 3\,MgF_2 + 2\,B + 3\,B_{25}O_{0.5}$$

$\Delta F°_{298}$ −522.6   −752.4

In these equations the Moissan boron is represented by $MgB_{25}O_{0.5}$ and the upgraded boron by $B_{25}O_{0.5}$. If it is assumed that the free energy of formation of Moissan boron is small, there would be a free energy decrease for each of these reactions. It seems likely, therefore, that MgF_2 is formed in the upgrading process.

Figure 1 shows electron micrographs of Moissan and upgraded boron taken at Picatinny Arsenal. Electron micrographs have indicated that particles varying from less than $0.1\,\mu$ to about $0.6\,\mu$ in diameter are present.

In the third step the 95-97% boron is converted to crystalline boron having a purity of 99%, or better, by heating in a vacuum to near or above the melting point. The 95-97% boron is cold-pressed, at 6000-7000 psi, into ring-shaped forms which can be stacked to form a hollow cylinder of boron resting on a boron disk. The inner

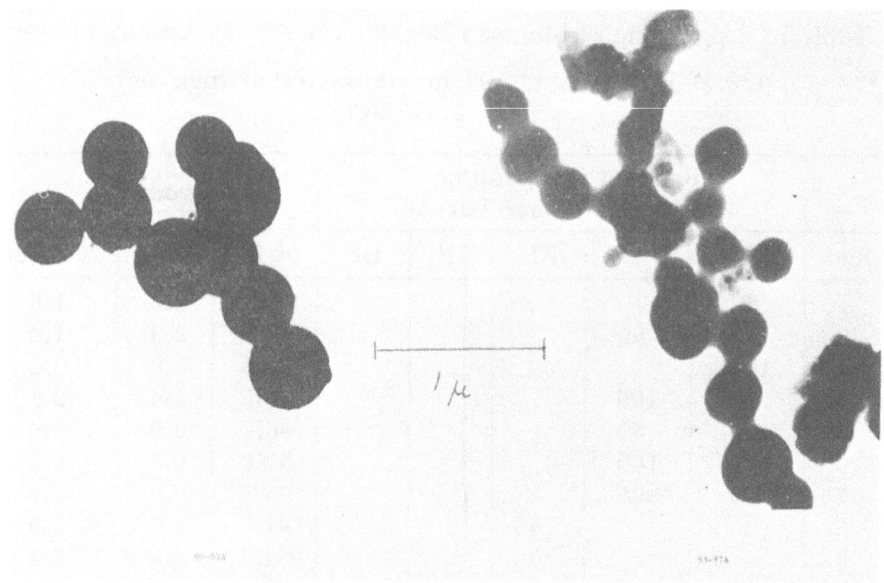

Fig. 1.

wall of this cylinder is heated by an inserted resistance heater. It is preferable to heat slowly enough that the pressure does not rise above 0.2 mm. After cooling, the portion of the wall next to the resistor is found to have fused and crystallized. In some cases crystal faces ½ in. or more long have been observed on the inner surface of the cylinder. Next to this fused layer is a thin layer of black material which has crystallized but not fused into a solid mass. Between this and the outside of the cylinder is a layer of apparently unchanged brown boron. These unfused portions of the cylinder are removed mechanically. The fused portion is then crushed in iron equipment, boiled with HCl to dissolve any iron, and then washed and dried. The purity of the product is generally over 99% and has been as high as 99.7%, with single crystals of 99.8%. The purity of the various raw materials and products is shown in Table II.

This table indicates that the main effect of the fluoride roasting is the removal of magnesium, which is converted into a compound soluble in hot acid, probably MgF_2. In some cases there was also a removal of impurities other than magnesium, which we believe to be mostly oxygen.

The literature indicates that in the Moissan process the maximum purity is obtained when the B_2O_3/Mg weight ratio is about 3:1. Our experiments confirmed that the 3:1 ratio can give a higher percentage of boron than the 2:1 ratio. However, the magnesium content is lower and the impurities other than magnesium (mostly oxygen) are somewhat higher in the 3:1 product.

Table II. Analysis of Raw Materials and Products

| | Raw materials | | Products | | | |
	Mg	B_2O_3	89% B	95-97% B	Crystalline	90-92% B
B			89.1	95.1	99.1	90.8
Mg	99.8		7.8	1.0	0.40	4.1
Al		0.05		0.01*	0.17	
Fe		0.03		0.24	0.11	0.13
Si	0.01 max	0.07		0.35	0.04	0.18
Mn	0.15 max			0.1*	0.10	
Cu	0.02 max			0.01*		
Pb	0.01 max					
Ni	0.001 max					
Ca				0.02*		
Cr				0.008*		
N			0.3	0.06-0.3	0.01-0.02	0.02
O†			2.4	3.1	0.06	4.6

*Spectrographic
†Estimated by difference

We found the 2:1 product to be upgraded more than the 3:1 product by the fluoride roasting procedure. The high magnesium content gives a product more subject to attack by the fluoride, and a greater fraction of the magnesium is removed. Table II also indicates that in the fusion process most of the remaining magnesium and practically all the oxygen are volatilized. The deposit on the walls of the vacuum chamber often contained reducing matter, suggesting that a lower oxide of boron was volatilized during fusion.

Although we believe the "filament" methods are more suitable for the preparation of extremely-high-purity boron, the present method is more convenient for preparing crystalline boron in quantity, and is probably capable of producing 99.7-99.9% boron. The chief impurities in the present 99.1% B product are Mg, Al, Fe, Si, and Mn, of which all but Mg could be largely eliminated by starting with purer boron oxide and magnesium.

VARIOUS PREPARATIONS OF ELEMENTAL BORON

Ray C. Ellis, Jr.*

Elemental boron has been prepared by the hydrogen reduction of boron bromide on hot wires and by other methods. Crystals up to 0.1 mm in length were grown on a tantalum wire at 1500°C. Boron diffusion into the wire at this temperature was serious. B_4C crystals up to 1 mm in diameter were grown on a graphite disk heated to 2000°C. Infrared transmission, resistivity, and hardness tests were made on the samples.

Several different methods of preparing boron single crystals have been used. This presentation summarizes this work, describes the crystals obtained, and reports the measurements that have been made.

It soon became evident from previous work [1,2] that somewhat larger crystals of boron would be desirable to facilitate various measurements and for possible device applications. The four methods of preparation attempted met with varying degrees of success.

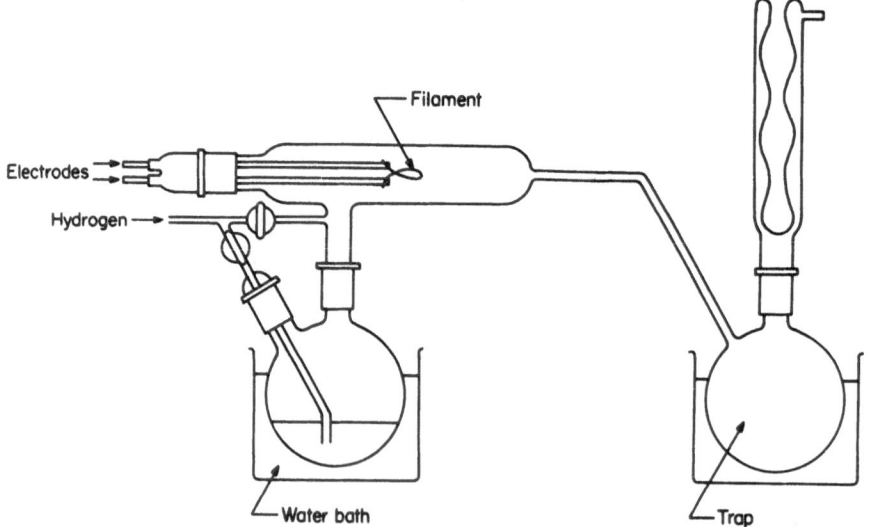

Fig. 1. Apparatus for the preparation of boron single crystals using a heated wire.

* Research Division, Raytheon Co., Waltham, Mass.

The first method attempted was essentially that of Laubengayer and co-workers [1]. The apparatus used is pictured in Fig. 1. The apparatus consisted of a heated wire filament held in an atmosphere of hydrogen in which a partial pressure of BBr_3 could be varied by the regulation of the temperature of a water bath. The unused BBr_3 and HBr byproducts were caught in a trap. Tungsten, molybdenum, and tantalum filaments were used, with tungsten showing the most resistance to boride formation.

The type of deposit depended upon the filament temperature. At 700°C the deposition rate was very slow, giving a smooth, microcrystalline deposit of boron. At about 900°C the deposit became lumpy, but was still microcrystalline. A photograph of such a deposit is shown in Fig. 2. At 1200°C crystals could be seen in the deposit. Figure 3 shows the deposit on a 20-mil wire. At temperatures of 1500°C and over, the wire invariably failed after a very short reaction time. After each experiment the wire was sectioned and a penetration of boron into the wire was observed. Typically, a

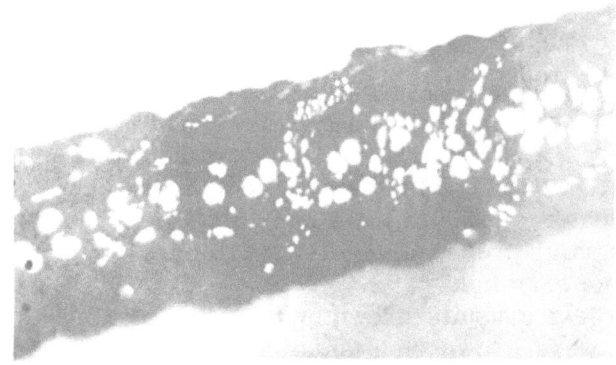

Fig. 2.

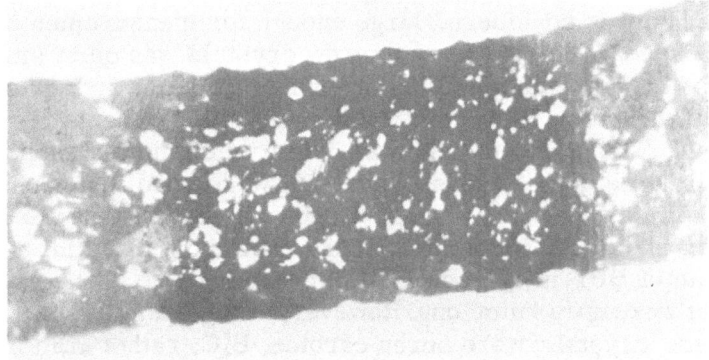

Fig. 3.

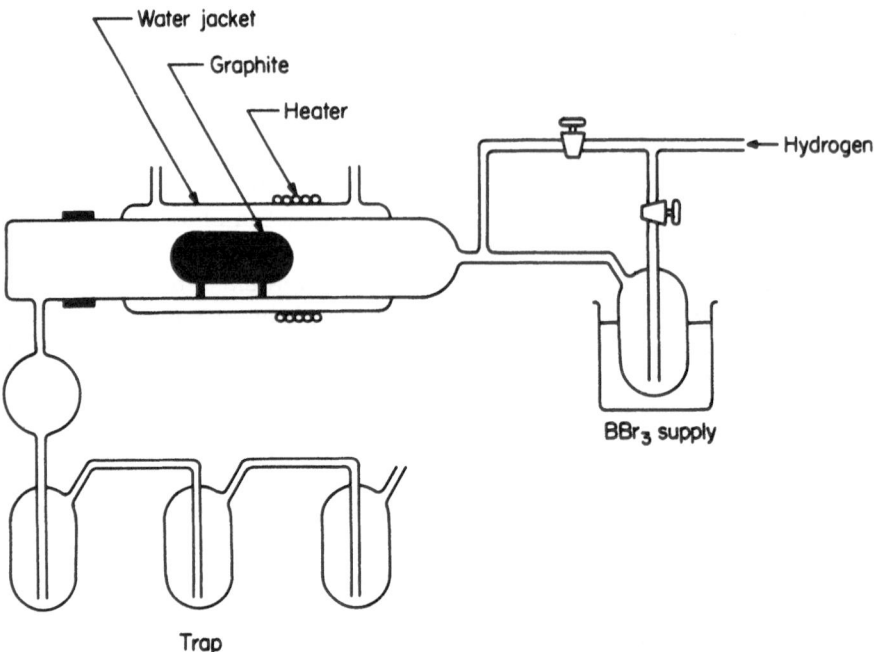

Fig. 4. Apparatus for the preparation of boron carbide single crystals using a heated graphite slug.

boron penetration of 1 mil was seen on a 20-mil tungsten wire that had been heated at 1200°C for 1 hr in a hydrogen atmosphere containing a partial pressure of BBr$_3$ of 19 mm. The tungsten boride layer was not examined.

The largest crystals grown by this method were about 0.1 mm long. They were grown on a tantalum wire at 1500°C in an atmosphere of H$_2$ containing a 60-mm partial pressure of BBr$_3$ in the 5 minutes before the wire failed. None of the crystals produced by this method was considered large enough for measurements.

The second method used to grow crystals was quite similar to the first except that an induction-heated graphite disk or slug replaced the filament. A schematic drawing of the apparatus is pictured in Fig. 4. The graphite slug was 1¼ in. in diameter, supported on short graphite legs. Tubes of graphite were used, but usually they failed after a few hours. Figure 5 shows a deposit that was formed in 56 hr at 1800°C in an atmosphere containing a partial pressure of 60 mm of BBr$_3$ in hydrogen. The largest of these dark gray, shiny crystals were up to 4 mm long. However, x-ray examination revealed that these crystals were boron carbide, B$_4$C, rather than boron. It is believed that the carbon was supplied to the growing crystals from the vapor. It is known that a partial pressure of methane is estab-

Fig. 5.

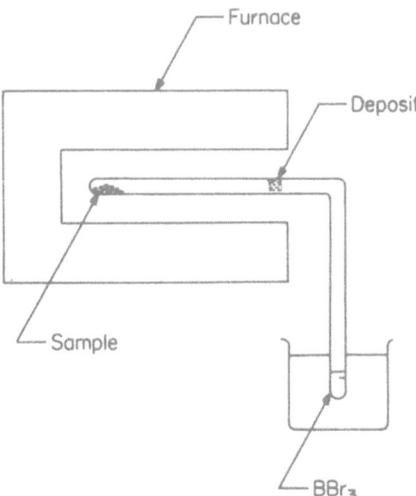

Fig. 6. Apparatus for the transport of boron using lower halides.

lished when carbon is heated in the presence of hydrogen [3]. It is very unlikely that the carbon was supplied to the crystals by diffusion from the substrate because of the relatively great distance.

Sample crystals were cut into 40-, 20-, and 10-mil thick disks. No transmission was observed from 2 to 20 μ using a Perkin-Elmer double-pass monochromator, with a microscope attachment. Boron carbide appears to be a semiconductor. The material pictured in Fig. 5 had a resistivity of 20-25 ohm-cm, p-type, with an energy gap of 0.15 ± 0.08 ev calculated from the slope of the log ρ vs $1/T$

curve. It was suggested by V. P. Jacobsmeyer that this measurement is in error because it was done in air rather than in vacuum. The hardness of this material was measured to be 2963 KHN with a 500 g load. Further study is needed on this interesting material.

Another method used to grow boron crystals utilized the relatively high vapor pressure of boron near the melting point. In a typical experiment, a charge of 150 g of boron, supplied by Pacific Coast Borax Company, was placed in a 1¾-in. ID graphite tube and heated at 2000°C for 24 hr in an argon atmosphere. Some boron evaporated and recrystallized as needles on cooler parts of the tube and boron charge. Needles up to 100 mils long by 1 mil in diameter were obtained. The crystals were very delicate and became beaded with clear B_2O_3 droplets when heated in air. These crystals were not investigated further.

The last method used to prepare boron crystals is based on a paper by Schäfer and Nickl [4]. They stated that boron, like silicon, may be transported from a hot zone to a cool zone by high-temperature, unstable, lower halides.

Accordingly, 1 g of boron and 1 g of BBr_3 were placed in a long, bent quartz tube. The tube was evacuated, and the end of the tube with the boron was placed in a furnace at 1000°C as pictured in Fig. 6. The other end was placed in a water bath and kept at room temperature. The total pressure in the system, then, was the vapor pressure of BBr_3 at room temperature.

Molecules of BBr_3 colliding with the elemental boron sample form a lower halide (or halides). This halide diffuses to a cooler portion (approximately 900°C) of the tube, where it decomposes. After 700 hr the tube was broken open and the deposit examined. Approximately 30% of the boron had been transported to a lumpy, microcrystalline mass which adhered to the quartz tube. This experiment yielded very impure boron because of the evidence that the product had reacted slowly with the quartz,

$$4B + 3SiO_2 \longrightarrow 3Si + 2B_2O_3.$$

A quantity of low-melting $B_2O_3 - SiO_2$ mixture appeared with the adhered boron and formed many small cracks during cooling.

An interesting experiment, demonstrating that this type of reaction might be progressive, can be performed easily. A piece of elemental boron is placed on a quartz plate in a furnace heated to 950°C in air. Figure 7 is a drawing of the reaction as it progresses. The boron does not react immediately with the air. The boric oxide layer in contact with the quartz slowly dissolves some silica and draws away, exposing the boron. This in turn oxidizes, forming more B_2O_3. The boron continues to oxidize until only the oxide ring remains.

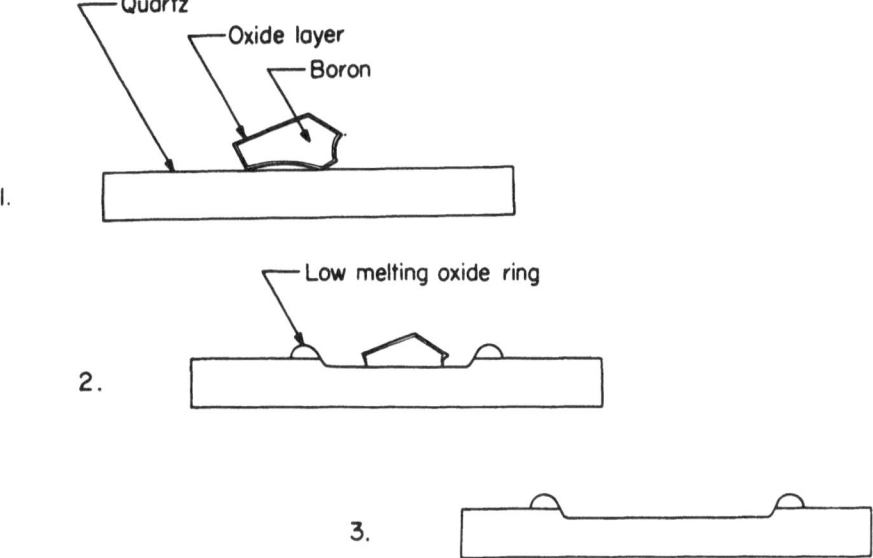

Fig. 7. Progressive oxidation of a boron chip in contact with quartz.

To summarize, four methods have been used in attempts to prepare boron crystals. Three of these yielded small or delicate, unusable crystals. The other method gave B_4C, an interesting semiconductor.

References

1. Laubengayer, A. W., Hurd, D. T., Newkirk, A. E., and Hoard, J. L., J. Am. Chem. Soc. 65 (1943) 1924.
2. Shaw, W. C., Hudson, D. E., and Danielson, G. C., Phys. Rev. 107 (1957) 419.
3. Pring, J. N., and Fairlie, D. M., J. Chem. Soc. 101 (1912) 91.
4. Schäfer, Harald, and Nickl, Julius, Z. anorg. allgem. Chem. 274 (1953) 250.

UTILIZATION OF BORON FILAMENTS IN VAPOR-PHASE DEPOSITION OF BORON

K. E. Bean and W. E. Medcalf[*]

An investigation has been made of several materials for use as filament substrates in the vapor-phase deposition of boron from BBr_3. Best results were obtained with boron and tantalum filaments. Boron crystal deposits 2 cm in diameter and 15 cm in length were deposited on small, vertical boron filaments.

Two types of boron filaments were investigated, including (a) small bars grown by the Czochralski process from boron nitride containers and (b) small-diameter float-zoned bars in which the original bar used in the zoning was a deposited crystal bar. The filaments grown by the Czochralski method normally contained trace impurities including Si, Fe, Mg, Ca, and Cu. These impurities in turn tended to be introduced into the subsequently deposited boron. Boron deposited on zoned bars (b) was entirely free of any impurities detectable by spectrographic analysis.

A method was developed in which crystal bars of boron of small diameter could be deposited on a straight tantalum filament in such a manner that the tantalum filament could be subsequently pulled from the deposit upon cooling, leaving the small crystal bar intact. The deposited boron bar was introduced directly into the floating-zone unit and zoned for any desired number of passes. Such zoned bars were used as filaments in subsequent vapor-phase depositions.

Studies of the cross sections of boron deposits on zoned boron filaments revealed no line of demarcation between filament and deposit except in those runs in which very dendritic boron was deposited. Both high-temperature (beta) and low-temperature (alpha) forms of the rhombohedral structure were deposited on boron filaments.

The vapor-phase deposition of boron was carried out by Laubengayer and co-workers [1] with deposition on iron, platinum, tungsten, and tantalum. They reported that iron and platinum were unsuitable, tungsten was satisfactory, and tantalum gave the best results. Stern

[*] The Eagle-Picher Research Laboratories, Miami, Oklahoma.

and Lynds [2] found molybdenum, tungsten, and tantalum to be unsatisfactory because the filaments became very brittle during the first hour of operation and the major contaminant in the boron came from the filaments. They found titanium filaments to be the most satisfactory since contamination due to titanium could be appreciably removed by chlorination at 300°C.

In the present study it was considered that a desirable material for use as an electrically heated filament for the deposition of boron should have the following characteristics:

1. relatively high current-carrying capacity
2. durability under the conditions of high temperature
3. resistance to destructive action on or by the deposited boron
4. inability to introduce impurities into the deposited boron
5. amenability to further processing such as crystal growing and float-zoning, without the necessity of removal from the deposit.

An experimental and critical study in this laboratory of many types of filament materials, including those mentioned above, resulted in the conclusion that boron filaments would most completely meet these requirements. It was further considered that hydrogen reduction of boron tribromide would be a desirable procedure in the production of high-purity boron since (a) boron tribromide reduction requires the lowest temperature of all the boron halides and (b) experience had shown that BBr_3 was more readily purified by fractional distillation than any of the other boron halides.

The preparation of boron filaments for use in the deposition of boron progressed through several stages. The most successful procedures were found to be

1. preparation of the filament by a combination of growing Czochralski crystals from a boron nitride container followed by float-zoning, and
2. deposition of a thin crystal bar of boron on tantalum in such a manner that the deposit could be slipped intact from the tantalum filament and then float-zoned to provide a strong, stable boron filament.

Experimental

The general experimental technique of carrying out boron deposition from BBr_3 in a hydrogen atmosphere differed but little from the general experimental technique known as the "filament method."

The essential parts of one form of apparatus used are shown in Fig. 1.

The reactor was made of clear fused silica with Teflon head plates. The reactor tube was connected to a fused-silica flask, used for storage of the previously purified boron tribromide. Connections in all cases were made with fused-silica ball-and-socket joints and

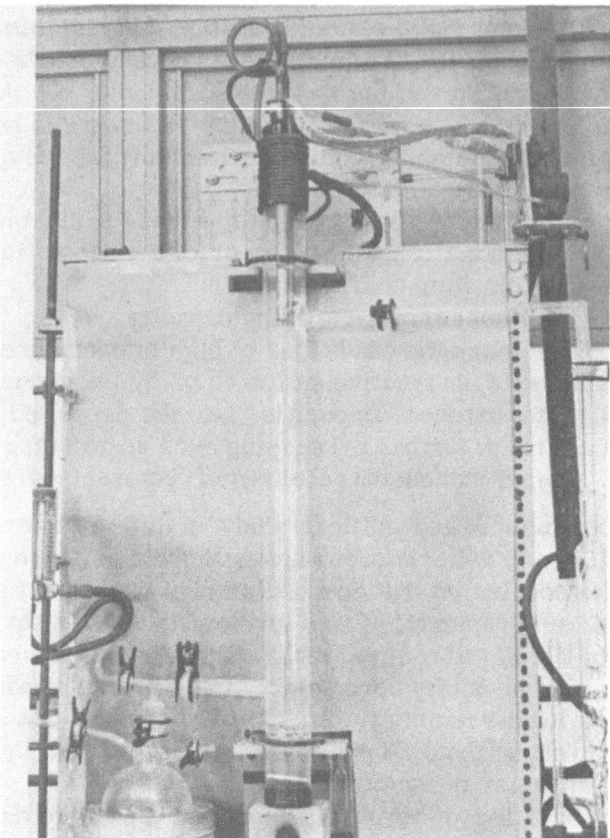

Fig. 1. Vitreosil unit used for decomposition of BBr₃.

were operated dry to avoid contamination with lubricant. The unit
shown in Fig. 1 contains a tantalum hairpin filament. Three different
electrode materials were used to hold the filament. They were (1)
graphite, (2) tantalum, and (3) Hastelloy B. A mixture of boron
tribromide and hydrogen entered at the bottom of the reactor and
exited near the top; the unreacted BBr₃ was condensed in the clear
Vitreosil unit which was attached to the reactor by fused-silica
ball-and-socket joints.

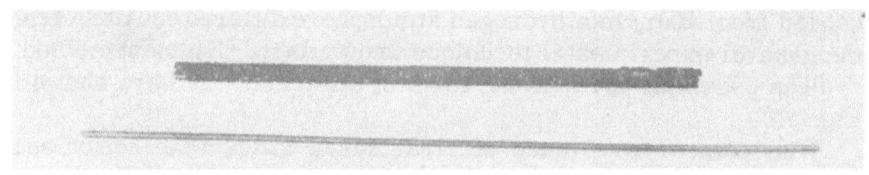

Fig. 2. Polycrystalline boron bar and tantalum filament from which
it was removed.

Boron can be readily broken from a tantalum filament by bending the deposit. In many cases, however, tantalum or tantalum boride chips off with the boron, causing contamination of the product. To eliminate this problem, a method was developed by which the crystalline boron bar could be slipped intact from the filament in such a manner that it is free of tantalum contamination. Figure 2 is a photograph of such a boron deposit that has been removed intact from the filament. The polycrystalline bar is 15 cm in length and 4 mm in diameter. The diameter of the longitudinal opening through the center of the bar is approximately 1.5 mm. The tantalum filament shown is the one on which the deposit was made and from which the boron bar was subsequently removed. The tantalum is not brittle and can be reused. Crystal bars such as these can be readily mounted in the floating-zone unit and zoned into strong, stable filaments. These zoned filaments can then be remounted in the reduction unit for use in the further deposition of boron.

In the development of a method which would allow deposition of boron crystal bars in a manner such as to allow them to be mechanically slipped from the tantalum filament, it was observed that tantalum boride (TaB_2) was formed at the junction of the tantalum and the boron deposit and that the thickness of the TaB_2 was a function of temperature and time. Formation of TaB_2 as a function of temperature is shown in Fig. 3. The photograph provides a comparison of cross sections of tantalum subjected to identical conditions of reactants and time, but varying in temperature. From left to right the temperatures were 800, 1175, 1200, 1250, and 1300°C. The temperature determination was made with an optical pyrometer. The tantalum boride deposit is visible on all filaments except the one on the extreme left. The filament second from left represents the approximate thickness of the tantalum boride layer normally found on filaments that can be mechanically slipped from the boron deposits

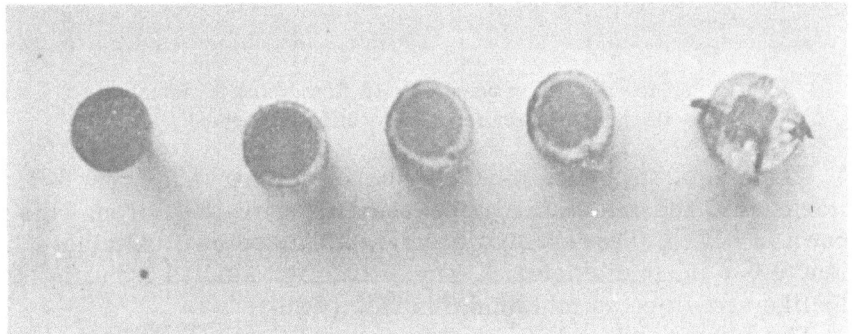

Fig. 3. Cross section of tantalum filaments after boron deposition.

without fracturing the crystalline material. Filaments with thicker deposits of TaB$_2$ could not be removed from the deposits without fracturing the crystalline boron bar.

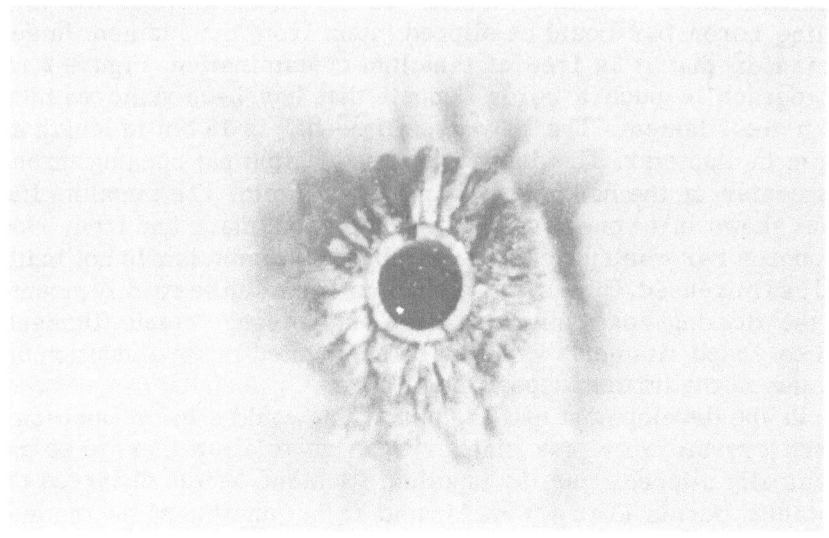

Fig. 4. Cross section of boron deposit from which the tantalum has been removed with HF.

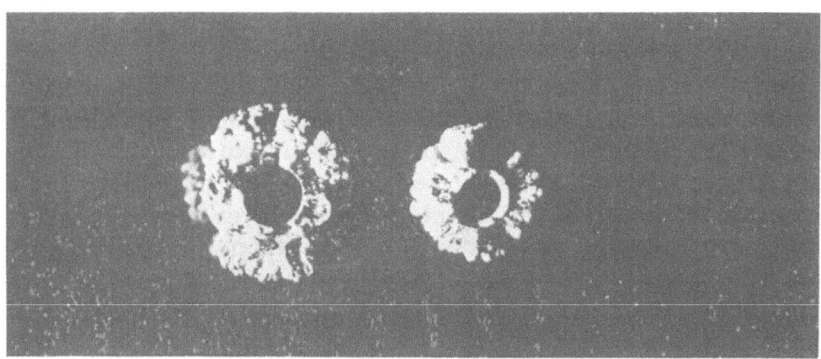

Fig. 5. Cross sections of boron deposits from which the tantalum filament has been removed by different methods.

When tantalum is leached from a boron deposit with warm hydrofluoric acid the tantalum boride remains with the boron. This is shown in Fig. 4. The deposit was originally made on a tantalum filament 0.060 in. in diameter. X-ray diffraction studies identified the tubelike structure as tantalum diboride (TaB$_2$).

The photograph in Fig. 5 compares a boron deposit from which the tantalum filament has been removed mechanically (left side of photograph) with a boron deposit from which the tantalum has been leached with hydrofluoric acid (right in photograph). It will be noted

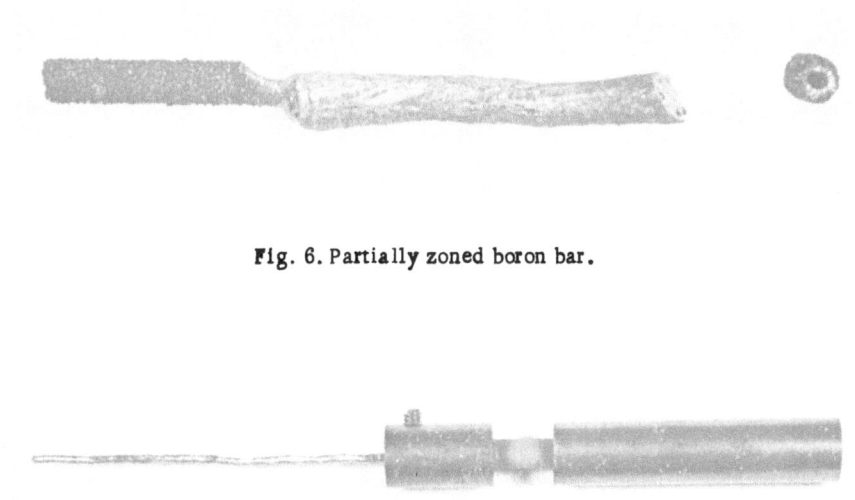

Fig. 6. Partially zoned boron bar.

Fig. 7. Zoned boron filament mounted in piston-type electrode.

that the tantalum boride is still present in the leached deposit but is not present in the boron crystal bar at the left of the photograph, providing evidence that tantalum boride is removed along with the tantalum filament when it is mechanically slipped from the crystalline boron bar.

The factors involved in depositing a crystal bar on tantalum in such a manner that it can be mechanically slipped from the filament obviously include temperature and time. Another factor is control of the molar ratios of H_2 and BBr_3 during the deposition. The successful procedure requires: (1) initial outgassing before deposition in helium to prevent hydriding of the filament; (2) initial deposition at low temperature (1100°C) for 30 min with a low mole ratio of BBr_3; (3) deposition at an increased temperature of 1150-1175°C for 1 hr with a high mole ratio of BBr_3. The deposit is then cooled quickly in an atmosphere of hydrogen.

Crystal bars slipped from the tantalum are mounted directly in the floating-zone unit and zoned to provide strength and current-carrying capacity. A partially zoned crystal bar is shown in Fig. 6. A cross section of the bar before zoning is shown at the right in the photograph.

A boron filament which has been completely zoned and mounted in an electrode is shown in Fig. 7. The filament is 13 cm in length. It was prepared by two zone passes in the floating-zone unit, using a crystal bar which had been mechanically slipped from a tantalum filament. It is mounted in a piston-type electrode ready for insertion in the boron tribromide decomposition unit. The piston-type elec-

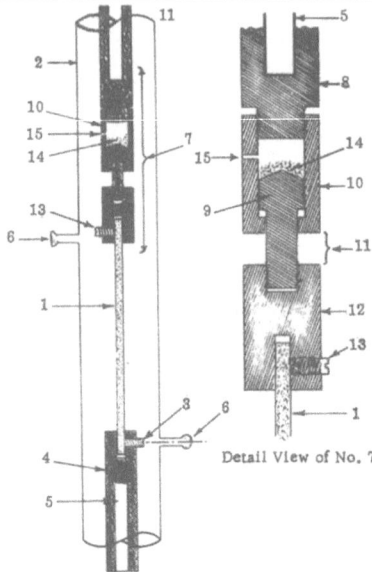

Fig. 8. Decomposition unit employing boron filament and piston-
type electrode.

1. Boron filament   8. Upper graphite electrode
2. Fused silica envelope   9. Graphite piston
3. Hastelloy B set screw on center   10. Graphite cylinder
   with gas orifice   11. Expansion zone
4. Graphite electrode   12. Graphite connector
5. H_2O cooling channel   13. Hastelloy B set screw
6. Halide gas orifice   14. Powdered graphite
7. Piston-type electrode   15. Gas escape port

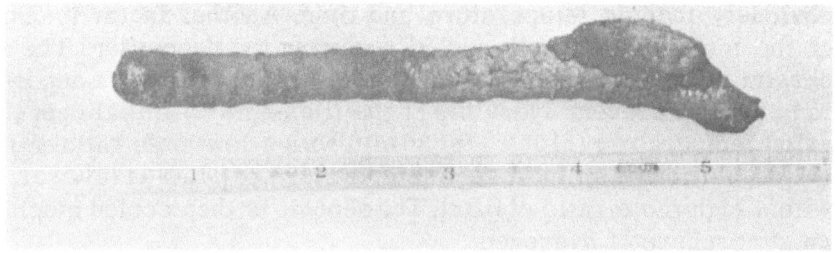

Fig. 9. Boron deposited on boron filament, piston-type electrode not
used (scale in inches).

trode is used to allow expansion of the boron filament during the
deposition.

Details of the piston-type electrode are shown in Fig. 8. The
electrodes are entirely of graphite. Cooling of the electrodes is
accomplished by bayonet-type water coolers inserted in the center
of each electrode.

Results

The effect of not providing an expansion zone during deposition
may be observed in Fig. 9. Deposition in this case was made on a

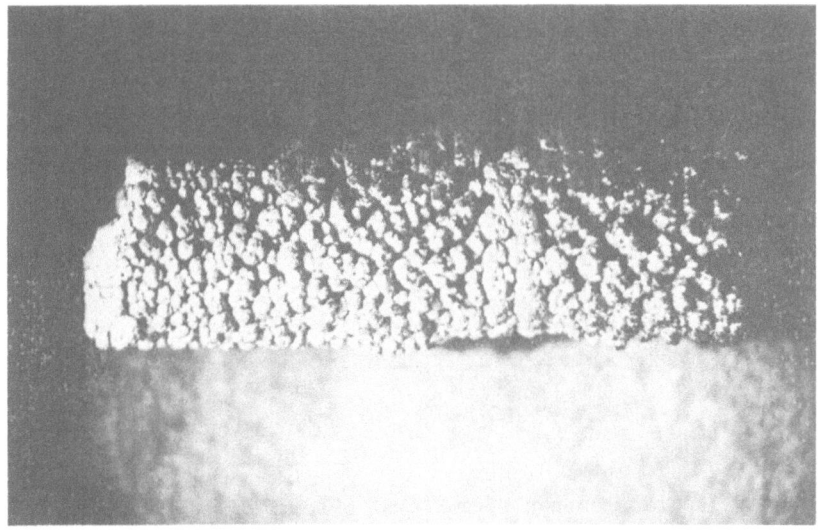

Fig. 10. Linear view of boron deposit on a small boron filament.

boron filament (the end of the filament can be detected at the extreme left in the photograph) but the piston-type electrode was not used. The linear expansion resulted in a bend in the filament (right, in photograph). A boron deposit of different appearance is noted at this end of the bar. The brown-appearing deposit consisted mostly of the low-temperature (alpha) rhombohedral form. The low temperature at this point (approximately 1000°C) was due to its proximity to the gas entrance of the reactor and the impingement of the entering cold mixture of H_2 and BBr_3 on the deposit.

A linear view of a crystalline boron deposit on a boron filament is shown in Fig. 10. The deposited boron is quite crystalline and is the high-temperature (beta) rhombohedral form. Boron deposited on boron filaments normally tends to be more uniform in size and in crystal structure than is the case when the deposition is made on more metallic filaments. Figure 11 is a photograph of a cross section of boron deposited on a boron filament. The filament was only $\frac{1}{16}$ in. in diameter and the deposit is $\frac{1}{2}$ in. in diameter. It will be noted that the center is a dense crystalline mass and there is no visible demarcation between the filament and the deposit. This dense-appearing boron in the center was deposited at approximately 1200°C and with a high concentration of BBr_3 (approximately 20%). There is a contrast in appearance with the boron near the outer portion of the deposit. The difference in structure and appearance is due to the fact that the outer portion was deposited while using a lower concentration of BBr_3 and a higher deposition temperature.

In Fig. 12 a line of demarcation between the boron filament and deposit can be clearly detected. Deposition in this case was made at a higher temperature (1280°C) and with a lower concentration of BBr_3 in the gaseous reaction mixture (10 mol. %). The deposit was

Fig. 11. Cross section of a dense deposit of boron on a boron filament.

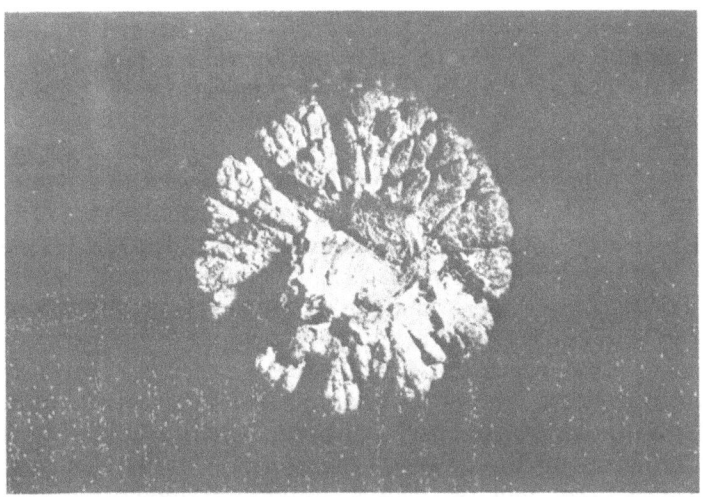

Fig. 12. Cross section of a dendritic deposit of boron on a
boron filament.

more dendritic and provided greater contrast to the filament in crystal structure and appearance.

The average conditions prevailing in runs carried out in the deposition of boron on boron filaments by the hydrogen reduction of BBr$_3$ are shown in Table I.

Table I. Average Conditions Prevailing in the Deposition of
Boron on a Boron Filament

Optimum temperature . . . . 1300°C
Optimum concentration of BBr_3 . . . 20 mol. %
Residence time . . . . approx. 10 sec
Hydrogen flow rate . . . . 2 ft^3/hr (970 cc/min)
BBr_3 volatilization rate . . . . 60 cc/min
Rate of deposition . . . . 2 to 3 g/filament/hr
Recovery as elemental boron. . . . . 35% of theoretical
Power input . . . . 65 amp at 20 v (1.3 kw)

The purity of the boron deposited on a boron substrate was as-
certained by (a) direct determination of the boron concentration by
volumetric determination as boric acid in the presence of mannitol
and (b) determination of the impurities present by spectrographic
and chemical methods. Approximately 20 determinations of boron
were made by the direct chemical method. The values ranged from
99.5 to 100.4%. Determinations on the same samples varied as much
as 0.4%.

Spectrographic determinations of impurities present are shown
in Table II.

Table II provides evidence of the following:

1. The boron filament prepared from a Czochralski crystal grown
   in a boron nitride crucible contains more impurities than does
   the filament which was prepared by float-zoning boron which
   had been mechanically slipped from a tantalum filament.
2. The deposited boron tends to pick up impurities from the substrate on which it is deposited, and the concentration of these
   impurities in the deposit decreases as a function of distance
   from the substrate.
3. The impurity most commonly found in all deposits is silicon;
   however, selected samples contain no spectrographically detectable impurities.

Conclusion

Crystal bars of boron can be deposited by the filament vapor-
phase method on tantalum filaments in a manner that allows them to
be mechanically removed intact without tantalum contamination.
Such bars can be float-zoned to provide strong, electrically conduct-
ing filaments. Boron deposited on such filaments is free of spectrographically detectable impurities.

Table II. Spectrographic Analytical Data on
Boron Deposited from the Vapor Phase

A. Deposited on boron filament prepared by growing from melt on boron nitride pedestal

Ref. No.	Description	Impurities (ppm)					
		Si	Mg	Fe	Cu	Ca	Al
M5906AO	Boron nitride pedestal	2000	50	10	5.0	1000	20
M5906AE	Boron filament (after deposition)	50	1	1	1.0	5	ND
M5906AC	Boron deposit next to filament	10	2	3	2.0	3	ND
M5906AB	Boron deposit farthest from filament	5	1	ND	1.0	1	ND

B. Deposited on boron filament prepared by zoning crystal bar slipped from Ta filament

M5908EP	Boron filament after deposition	5	ND	ND	ND	ND	ND
M5908EQ	Boron deposited on filament	10	ND	ND	ND	ND	ND

C. Selected samples of boron deposits

M5905DZ		ND	ND	ND	ND	ND	ND
M5905EA		ND	ND	ND	ND	ND	ND

D. Red modification -- low-temperature (alpha) rhombohedral

M5908EW		50	1	ND	0.5	ND	ND
M5908EX		50	5	ND	ND	ND	ND

ND = Not detected

Spectrographic detection limits (ppm): Si 1, Mg 0.5, Pb 3, Cu 0.5, Ag 0.5, Ca 0.5, Fe 1, Sn 2, In 5, Al 2, Zn 50, Ta 500

Acknowledgment

Acknowledgment is given to the U.S. Army Signal Research and Development Laboratory for their sponsorship of this work.

References

1. Laubengayer, A. W., Hurd, D. T., Newkirk, A. E., and Hoard, J. L., J. Am. Chem. Soc. 65 (1943) 1925.
2. Stern, D. R. and Lynds, L., J. Electrochem. Soc. 105 (1958) 676.

GROWTH OF BORON CRYSTALS BY THE CZOCHRALSKI AND FLOATING-ZONE METHODS

R. J. Starks and W. E. Medcalf[*]

An investigation has been made of various ways to densify and grow crystals of boron. Induction heating methods were used to densify the boron in boron nitride crucibles and on pedestals of boron and boron nitride. Crystals approximately 6 in. in length were grown by the Czochralski method from densified charges.

Floating-zone studies were carried out on bars prepared by (a) the Czochralski growth process from boron nitride crucibles and (b) crystal bars as deposited in the vapor phase deposition. Five zone passes were normally made, with the liquid zone moving from the top to the bottom of the vertical ingot. The zoned bars grown from boron nitride crucibles by the Czochralski process tended to crack with zoning after one or two zone passes.

Spectrographic analysis of the zoned bars showed that Si, Fe, Cu, V, and Ca concentrated in the bottom end of the zoned bar, indicating that these impurities had segregation coefficients less than unity. The concentration of magnesium decreased toward the lower end of the bar. Electrical measurements made on the zoned bars showed that the resistivity ranged from 10^7 to 10^4 ohm-cm, decreasing toward the bottom portion of the bar. Small monocrystalline areas were found in the float-zoned bars which x-ray studies showed to be of the rhombohedral structure. Comparison is made of properties of the bars zoned in vacuum and ambient gases.

The primary purpose of this investigation has been to find a way to produce massive crystalline ingots from boron prepared by the hydrogen reduction of boron tribromide. The boron available for densification and crystal growth has been in four forms (Fig. 1): (1) small pieces of the deposit broken from the substrate, (2) cylindrical pieces of the deposit slipped from the substrate, (3) the deposited boron still intact on a tantalum substrate, and (4) the deposited boron still intact on a boron filament.

Methods have been developed to densify all these forms into massive crystals using induction heating techniques. Argon and

---

[*]The Eagle-Picher Research Laboratories, Miami, Oklahoma.

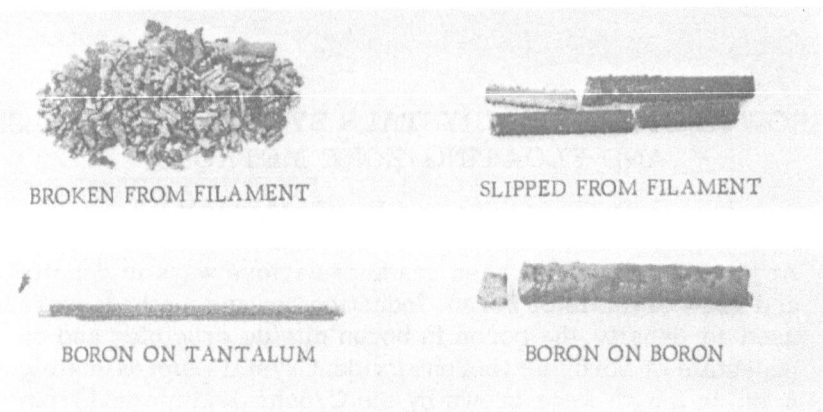

Fig. 1. Four forms of boron deposit prepared for densification and crystal growth.

Fig. 2. Boron ingot prepared by melting boron in a boron nitride boat (scale in inches).

helium have each been used as an ambient; crystal growth in a partial vacuum has also been studied.

Czochralski Crystals

Four different methods have been used to densify the small pieces broken from the filament as shown in Fig. 1.

(1) Bars can be prepared by melting the broken pieces of deposit in a cylindrical boron nitride boat inside a graphite heater (Fig. 2).

Fig. 3. Boron nitride crucible and heater used to melt small pieces of boron.

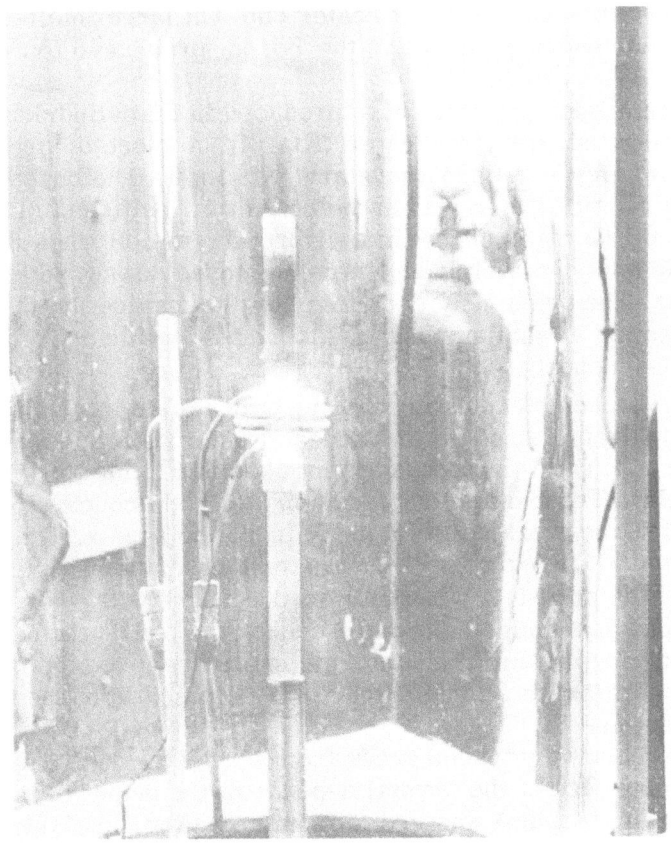

Fig. 4. Boron nitride boat being passed through a 5-Mc
induction heating coil.

The graphite heater was positioned in the center of a 2-in.-diam quartz tube mounted in a vertical position inside the work coil of a 450-kc induction heater. A zirconia block was used to support the heater and granules of zirconia were packed around the heater to protect the quartz tube. The boron nitride boats were ¼ in. ID and 3 in. long. During melting, the temperature as measured with an optical pyrometer was 2100°C. The melted boron coalesced and formed a dense bar near the top of the boat. As can be seen in Fig. 2, there was no apparent damage to the inner surface of the boron nitride tube, but the boron did "wet" the tubes near the top, and the tubes had to be sawed from the billets.

(2) Charges up to 12 g of these small pieces of deposit were melted in a standard "crucible form" made of boron nitride (Fig. 3).

The graphite heater cup (B) was machined so that the crucible (A) would fit snugly. The same quartz tube and coil was used as with the cylindrical boats. To protect the quartz tube, a boron nitride shield (C) was placed around the heater cup. The larger melts showed an increased tendency to "wet" the boron nitride and to crack on cooling.

(3) Crystalline bars were also produced in the cylindrical boats without using the graphite heater (Fig. 4). A piece of previously densified material is used to "start" the melt. The boat with the previously densified material in the bottom is positioned in a 5-Mc induction coil so that it may be moved through the coil when the melt forms. The boat shown in Fig. 4 is being moved downward at a rate of 5 in./hr. The previously densified lump was raised to the Curie point by heating a tantalum disk positioned between the boron nitride tube and the support.

(4) The broken pieces of deposit can be densified into solid crystalline lumps by melting them on a pedestal of boron nitride. Figure 5 shows the coil and work arrangement used in making these uncontained melts. The charge is placed on a slightly concave pedestal supported in the work coil of a 5-Mc induction heater. The charge is raised to the Curie point by heating a turn of tantalum wire around the base of the pedestal. When molten, the charge will coalesce into a nearly spherical lump. If allowed to cool on the pedestal, the charge will crack. To prevent cracking, the melts are grown onto a seed by the Czochralski technique. Figure 6 is a schematic of the furnace arrangement used.

Figure 7 shows a crystal grown from a pedestal-supported melt. The maximum size of the crystal depends on the output of the furnace used. This crystal was grown from a pedestal in a 1-in.-diam work coil; it weighs 4 g. The entire charge was grown from the pedestal. Normal freezing takes place until the charge is approximately three-fourths grown, at which time the "heel" will freeze to

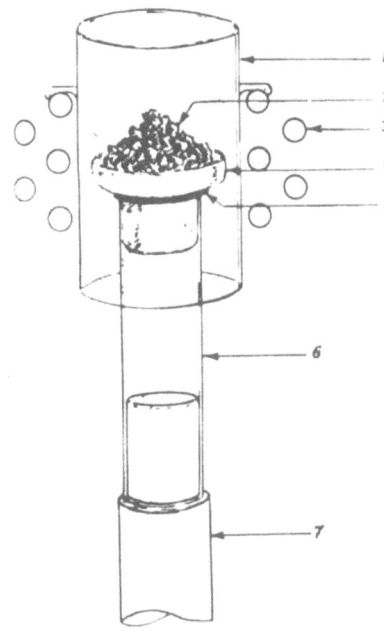

Fig. 5. Coil and work arrangement used in making uncontained melts. 1) quartz shield; 2) boron charge; 3) 5-Mc work coil; 4) BN pedestal; 5) Ta susceptor ring; 6) quartz support; 7) steel support.

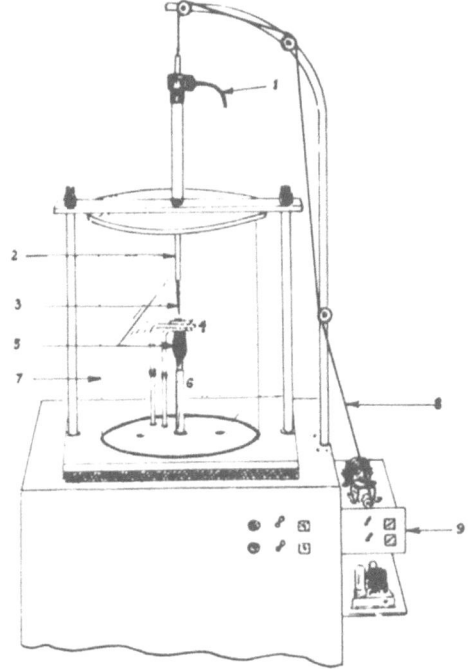

Fig. 6. Schematic of crystal furnace. 1) rotation drive; 2) stainless steel shaft; 3) seed; 4) 5-Mc work coil; 5) pedestal; 6) stainless steel shaft; 7) pyrex cylinder; 8) pull cable; 9) translation and rotation controls

Fig. 7. Crystal grown by Czochralski method from boron nitride pedestal.

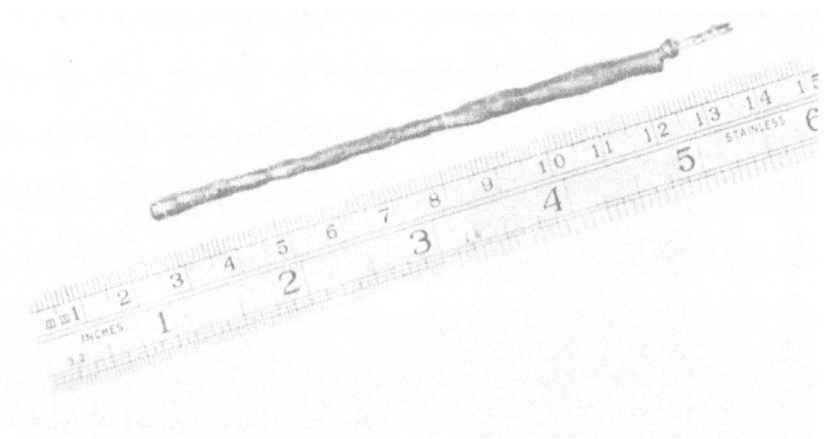

Fig. 8. Czochralski-grown crystal after one floating-zone pass.

the grown crystal. The shape of the crystal can be varied by changing the temperature and growth rate.

Czochralski-grown crystals of suitable diameter can be refined by float-zoning. If additional length is required, the grown crystal can be used to seed additional melts. Figure 8 shows a bar, prepared by growing two melts from a pedestal, after one zone pass.

Floating-Zone Crystals

The same 5-Mc furnace used for growing the Czochralski crystals, with a "pancake" work coil, is used for the float-zoning. The flat

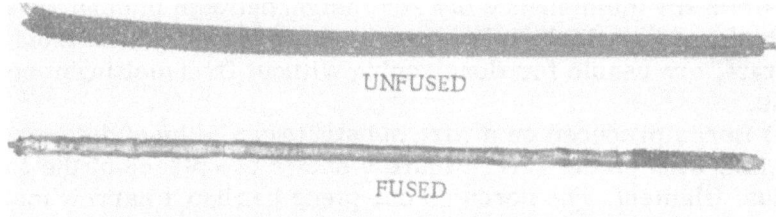

UNFUSED

FUSED

Fig. 9. Two pieces of the same tantalum filament with boron deposit: one unfused, the other fused by passing through the float-zoner.

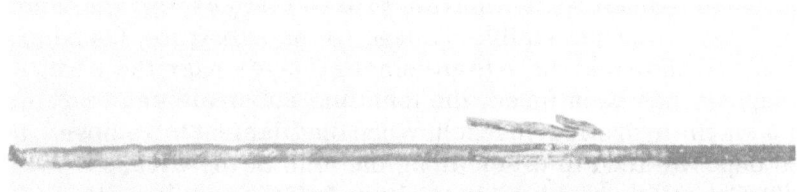

Fig. 10. Fused deposit on tantalum, broken to show undamaged tantalum.

Fig. 11. Partially zoned deposit that has been slipped from the tantalum substrate.

Fig. 12. Zoned bar of boron deposited on a boron substrate (scale in inches).

coil allows the maintenance of a zone height between 0.5 and 1.0 cm.

The forms available, other than the small pieces broken from the substrate, are usable for float-zoning without first melting in boron nitride.

(1) Boron produced on a wire substrate can be fused into a solid bar while still on the wire. Figure 9 shows two pieces of the same tantalum filament. The boron on one piece has had a narrow molten zone passed rapidly through the deposit. Careful adjustment of the temperature makes it possible to fuse only the outer portion of the deposit. The wire should be supported at each end to lessen movement that will cause the fused boron to crack. Figure 10 shows the same fused deposit with a portion broken off to expose the tantalum wire. There was no visible damage to the substrate. Close examination will show that there is an unmelted layer near the wire. After the deposit has been fused, the tantalum substrate can be removed with a warm hydrofluoric leach. When the filament is removed, these fused deposits tend to crack along the axis of the wire.

(2) Deposits which have been slipped off the tantalum filament are easily zoned. They are sufficiently strong to allow normal handling while mounting in the furnace. A turn of tungsten or tantalum can be used as a susceptor near the lower end of the bar. The bar shown in Fig. 11 has been partially zoned (one pass). One end has been broken off and turned to show the cross section of the starting bar.

(3) Boron deposits on zoned boron substrates are ideally suited for further zoning. Figure 12 shows such a bar after four zone passes.

Spectrographic analysis of zoned bars, both doped and undoped, have shown that most of the more commonly detected impurities have a segregation coefficient less than unity. Of the elements listed in Table I, only magnesium was present in smaller concentrations in the last end to freeze. This decrease in magnesium could be due to vaporization.

Table I.   Distribution Coefficients for
Impurities in Boron

Element	K
Silicon	<1
Iron	<1
Copper	<1
Vanadium	<1
Calcium	<1
Magnesium	>1

Spectrographic analyses of bars zoned in a vacuum compared with bars zoned in argon or helium have not been completed, but

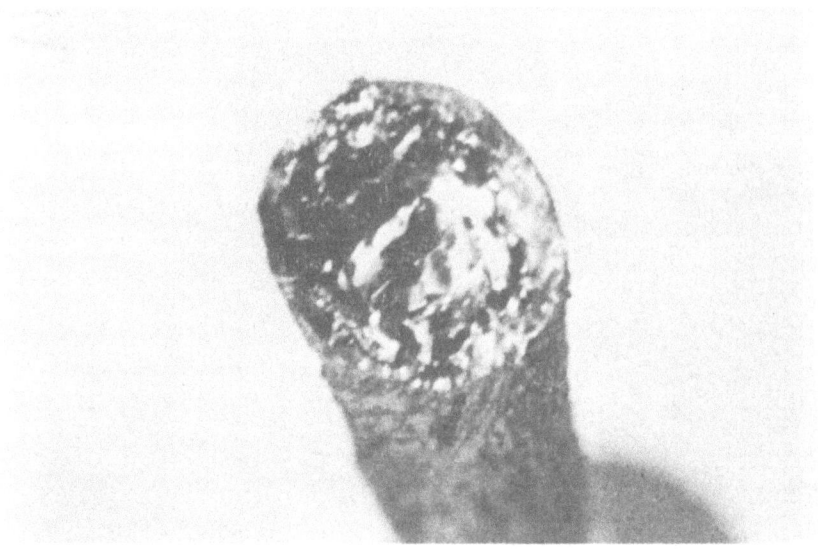

Fig. 13. Cross section of zoned bar prepared by slipping the deposit
from the substrate.

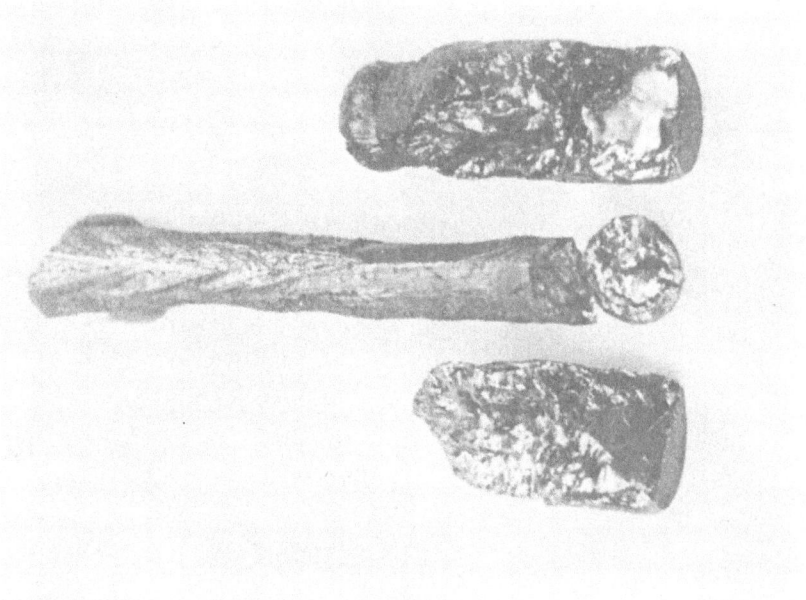

Fig. 14. Portions of zoned bars not prepared by melting in boron nitride.

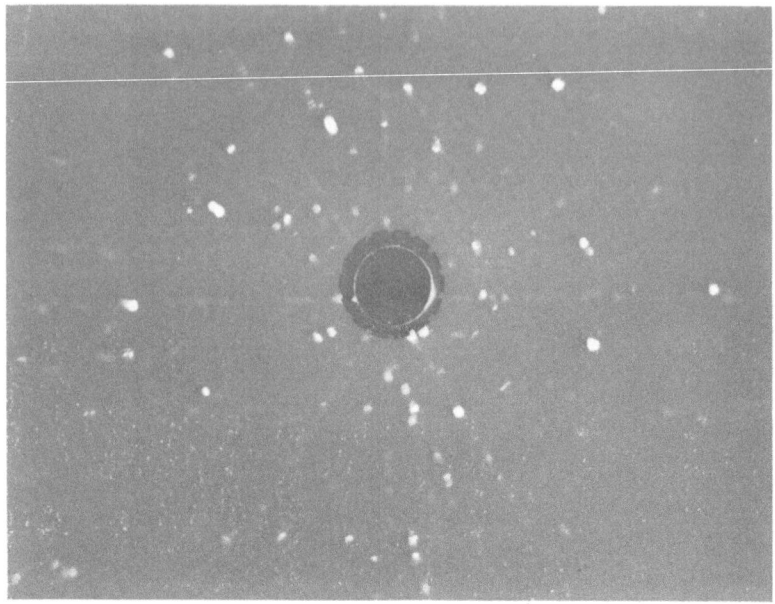

Fig. 15. Laue back-reflection x-ray pattern of boron crystal.

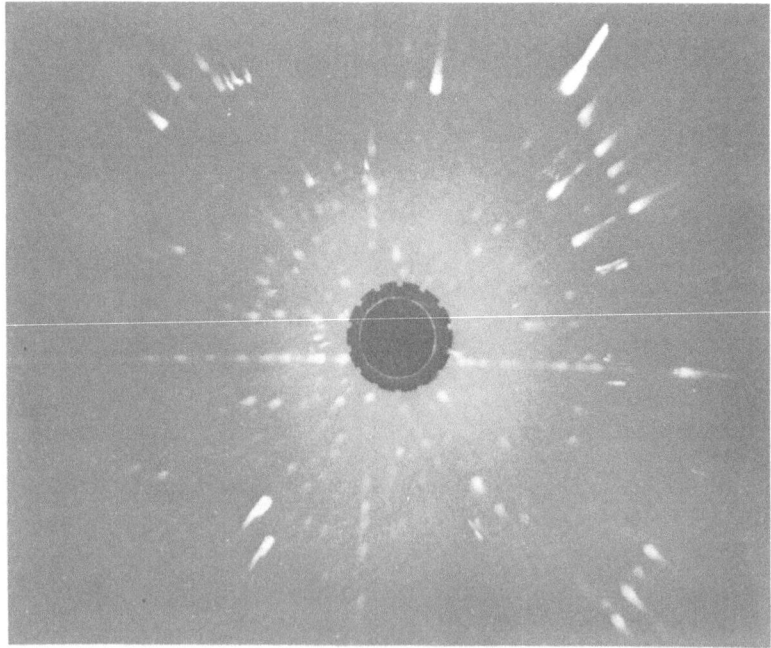

Fig. 16. Laue back-reflection x-ray pattern from end of bar zoned at 0.3 in/hr.

resistivity measurements indicate lower carrier concentration in the vacuum-zone bars. A bar that had been slipped from the filament was divided into two pieces for zoning. One zone pass on the bar in argon gave a resistivity of $1.15 \cdot 10^6$ ohm-cm, while one pass in a vacuum on the same starting material gave $1.72 \cdot 10^6$ ohm-cm.

A marked difference has been noted in the crystalline structure of zoned bars prepared from material that has been melted in boron nitride and that prepared by other methods. All bars prepared in boron nitride have shown a coring effect and tended to crack in a plane parallel to the growth axis. This condition has not been observed in bars prepared by slipping the deposit from the substrate or bars zoned on the boron filaments. Figure 13 shows the cross section of a zoned bar that was prepared for zoning by slipping the deposit from the substrate. There is no indication of coring in this bar. The center of the bar has the appearance of monocrystalline material. There is a thin layer of peripheral crystallites around the bar.

The center bar in Fig. 14 is a bar that had been slipped from the filament. It has had one zone pass. No coring was noted in this bar. The other pieces shown in Fig. 14 were broken from the tip of a zoned bar prepared by zoning a deposit still intact on a boron filament. The area near the right end of the bar appeared to be monocrystalline. A Laue back-reflection x-ray pattern made perpendicular to the plane of this face is shown in Fig. 15. Other x-ray patterns of this area show it to be largely monocrystalline, except for a band of peripheral crystallites.

The growth of these larger monocrystalline areas is also partially dependent on the growth rate. A bar ½ in. in diameter was seeded and grown at 0.3 in./hr and examined by x-ray techniques. Except for the layer of peripheral crystallites, the bar appeared to be monocrystalline. Identical Laue patterns were obtained from each end of the billet. Figure 16 shows this pattern.

Summary and Conclusions

(1) Large monocrystalline areas can be produced in zoned bars that have been prepared by slipping from the filament or deposited on boron filaments.

(2) Slower growth rates aid in producing large crystallites.

(3) Certain impurities can be segregated by the floating-zone method.

(4) Zoning in a vacuum gives greater resistivities than zoning in an ambient.

Acknowledgment

This investigation has been supported by Exploratory Research Division E of the Army Signal Research and Development Laboratory, Fort Monmouth, New Jersey.

ZONE PURIFICATION OF BORON

F. Hubbard Horn[*]

Boron has been zone-refined using boron nitride boats. Segregation coefficients for a number of impurities are known qualitatively. Crystalline boron has been grown by the Czochralski method. Some electrical, optical, and thermal data are discussed.

Elemental boron in the purity range 99.1-99.5% boron, obtained from several sources, has been zone-refined in an effort to obtain boron of the purity required for semiconductor investigations and to determine those impurities that cannot be removed readily by a crystallization procedure.

Hot-pressed hexagonal boron nitride was used as boat construction material. The system used to conduct the zone refining is shown schematically with dimensions in Fig. 1. It was found necessary to fire all parts made of boron nitride at about 1500°C in pure dry hydrogen for several hours. This procedure greatly reduced those

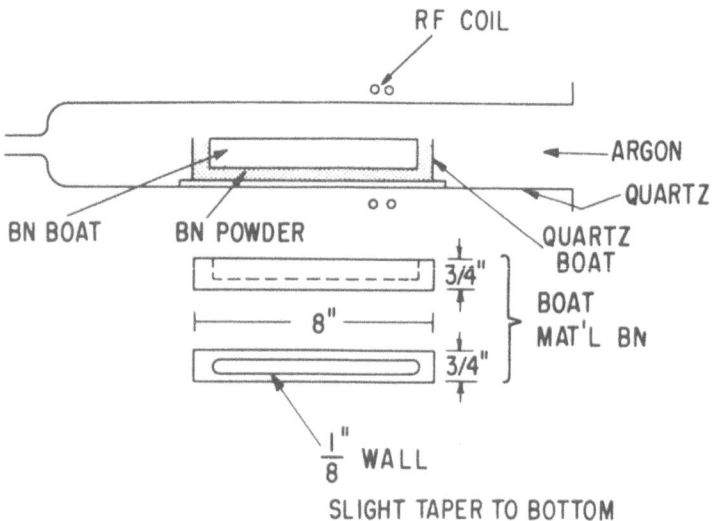

Fig. 1. System for zone refining boron (schematic).

[*]General Electric Research Laboratory, Schenectady, New York.

impurities which are volatile at the temperature of molten boron—
mainly boron oxides—and prevented sticking of the boron to the boron
nitride. Argon or helium gas was used as a protective atmosphere.
Heating was accomplished by means of induction heating at 450 kc.
At a temperature above about 1200°C boron will couple if the pieces
are large enough. Since our boron was in sizes from small granules
to chunks, it was necessary to melt the charge together in the boat
before attempting to zone refine. Some, but not severe, cracking of
the boron occurred during cooling. Solid pieces could be cut from
the ingots by use of a diamond saw.

X-ray emission analyses (combustion for carbon) of the original
boron and boron from the front, middle, and sprout portions of zone-
refined bars were made following an arbitrary number of zone passes
and on material from multiple generations of middle fractions. From
these data, an attempt was made to assess the segregation coefficients for impurities present in the original boron. These are given
in Table I.

Table I.   Impurity Segregation in Boron

$K > 1$	$K < 1$
(front of boat)	(sprout)
Al $\gg 1$	Fe $\ll 1$
C $\gg 1$	Ni $\ll 1$
Ti > 1	Cr $\ll 1$
	Cu < 1
	Si < 1

Magnesium, which was also present in the boron nitride (with traces
of iron and silicon), was not found in the zone-refined boron. It is
assumed that the magnesium was lost by volatilization.

The data for the segregation of impurities suggest that Si, Cu,
and Ti are not effectively removed from boron by crystallization.
These impurities should, therefore, not be present in boron prepared
chemically for semiconductor study.

The optical absorption was measured for zone-refined boron for
which an analysis of impurities had been made. All commercially
purchased boron had an absorption coefficient greater than 10^3 cm^{-1}.
In Fig. 2 are shown two typical optical absorption curves for zone-
refined boron. In A the total impurity content is about 0.1% (Cu plus
traces Ni and Fe) following five to eight zone passes using boron
of 99% purity. Curve B is for a section of boron with five zone passes
using boron of 99.6% purity. The analytical results show "traces".
It has been estimated from this and other optical absorption data
that curve B is for boron containing about 10^{18} carriers/cm^3.

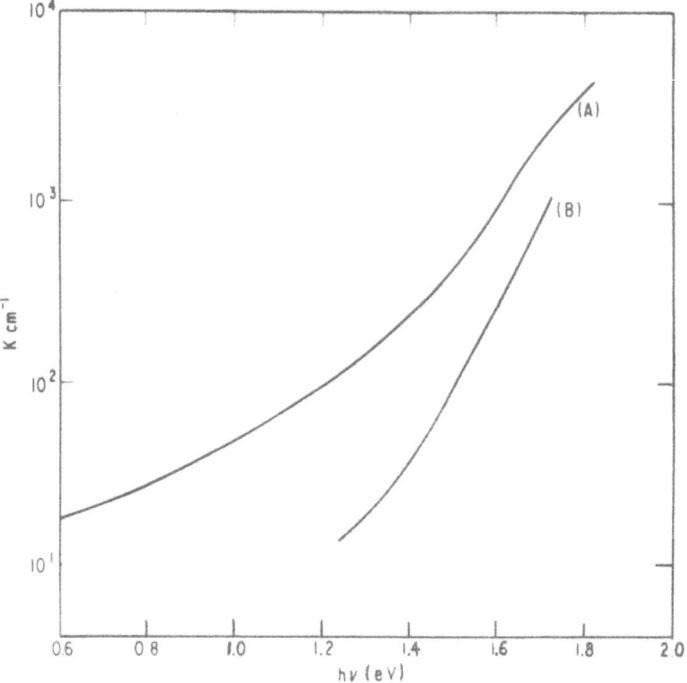

Fig. 2.

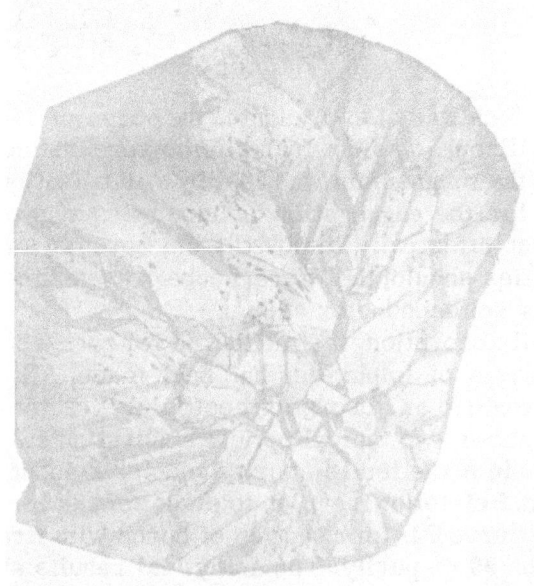

Fig. 3.

Figure 3 shows an infrared transmission photograph of a section of boron B above. In addition to the polycrystalline structure, numerous specks can be seen. It is not known whether these specks represent electrically active or inactive material that may have precipitated from the boron.

The electrical resistivity of boron B, containing approximately 10^{18} carriers/cm^3 is shown as a function of $1/T$ in Fig. 4. These data are in substantial agreement with the data of Greiner and Gutowski [1] for boron of similar purity. It appears likely that much of the

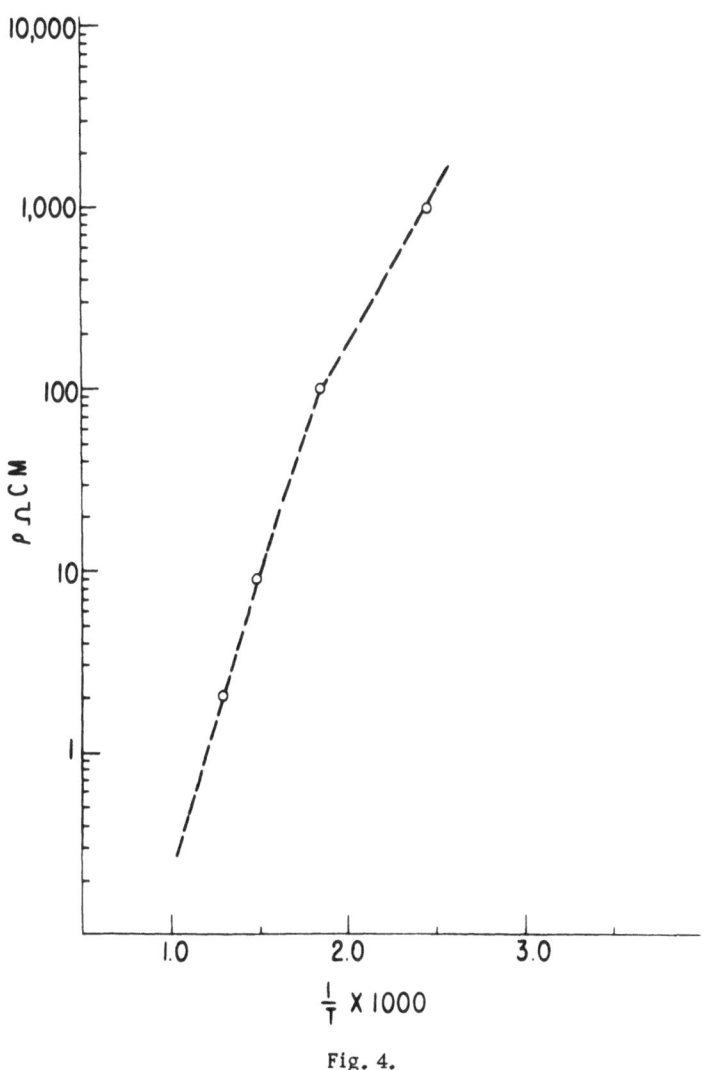

Fig. 4.

data for the electrical resistivity of boron is for boron in which the properties are masked by the presence of deep-level impurities.

All of the boron processed by melting has the complex rhombohedral crystal structure reported by Sands and Hoard [2], for which x-ray diffraction data have been published [3].

I wish to acknowledge that various portions of this report are based on the work of L. V. McCarty, R. O. Carlson, E. A. Taft, and W. C. Dash. Figures used in this report appear in the Journal of Applied Physics and the author wishes to thank the Editor for permission to use them.

References

1. Greiner, E. S. and Gutowski, J. A., J. Appl. Phys. 28 (1957) 1364.
2. Sands, D. E. and Hoard, J. L., J. Am. Chem. Soc. 79 (1957) 5582.
3. Horn, F. H., J. Appl. Phys. 30 (1959) 1612.

CRYSTALLOGRAPHY OF THE ALUMINUM BORIDES

J. A. Kohn*

At present, there are five authenticated phases in the aluminum—boron system:

1. AlB_2—hexagonal; bronze-colored, thin platelets.
2. AlB_{10}—orthorhombic; black lustrous, pyramidal and tabular; new phase.
3. α-AlB_{12}—tetragonal, pseudocubic; hematite-like plates and laths (irregular pseudooctahedra); thin sections orange-red in strong transmitted light.
4. β-AlB_{12}—orthorhombic, pseudotetragonal; amber-colored, bipyramidal; polysynthetically twinned on (110) and $(1\bar{1}0)$.
5. γ-AlB_{12}—orthorhombic, polytypically related to α-AlB_{12}; hematite-like, orthogonal laths; new phase.

(1) is the only aluminum boride phase whose properties are different from those of boron and also the only phase whose structure is known. (3), (4), and (5) have common unit cell vectors which show that, as in boron carbide and the known forms of boron, boron icosahedra represent the major structural elements.

Introduction

In general, the aluminum boride family of compounds bears a striking resemblance, regarding both structure and properties, to the modifications of elemental boron. This is pointed up by the fact that certain aluminum borides were actually considered for many years to be phases of boron, and only recently have they been chemically and crystallographically characterized.

With the exception of one compound (AlB_2), the aluminum borides are high-melting, extremely hard, relatively refractory materials. They offer potential for high-temperature (say, up to 700°C) electrical and optical application [1]. It is with this in mind that several industrial and government research laboratories are now actively interested in the aluminum borides. A few properties have been determined, one of which is described elsewhere in these Proceedings [2]. The present discussion deals with the crystallographic characterization of the aluminum borides and the relationships among them.

---

* U.S. Army Signal Research and Development Laboratories, Fort Monmouth, New Jersey.

Table I. Summary of Aluminum Boride Data

Property \ Phase	AlB_2	AlB_{10} (4)	α-AlB_{12} (4)	β-AlB_{12} (4)	γ-AlB_{12}
Color	Bronze	Lustrous black	Hematite-like Red in trans. lt.	Amber	Hematite-like Red in trans. lt.
Habit	Hexagonal platelets	Pyramidal Tabular	Pseudooctahedral, platy	Bipyramidal	Laths Orthog. plates
Sym. class	Hexagonal	Orthorhombic	Tetragonal, pseudocubic	Orthorhombic, pseudotetrag.	Orthorhombic
a, Å	3.009	8.88_1	10.16_1	12.34	16.6
b, Å	–	9.10_0	–	12.63_1	17.5
c, Å	3.262	5.69_0	14.28_3	10.16_1	10.2
Z	1	5.2_0	$14._5$	16	29
Sp. Gp.	$P\,6/mmm$	$B\,2/b\;2_1/m\;2/m$	$P\,4_2\,2_1\,2$ (?)	$I\,2/m\;2/m\;2/a$	$P\,2_1\,2_1\,2_1$
Density, g/cm^3	(3.17)	2.53_7	2.55_7	2.60_0	(2.56)
Twinning, etc.	–	–	Twinning, (101)	Polysynthetic twinning (110) & ($1\bar{1}0$)	Syntactic inter-growth with α

The Known Aluminum Boride Phases

The present complement of established aluminum boride phases numbers five structures. Physical data pertinent to these phases are summarized in Table I.

(a) AlB_2 — This "lower boride" of aluminum is the one atypical phase alluded to above. Crystals occur as soft, bronze-colored hexagonal platelets. The phase has a simple hexagonal structure described by Hofmann and Jäniche in 1935 [3]. It is totally unlike boron with respect to structure and properties. At present there appears to be no practical interest in AlB_2.

(b) AlB_{10} — Along with the three established AlB_{12} phases, AlB_{10} is one of the so-called "higher borides" of aluminum. Recently discovered at this laboratory [4], it has been obtained in single crystals up to 2 mm in largest dimension. Several typical crystals are shown in Fig. 1; both pyramidal and tabular habits are observed. The pyramid axis is [010]; the table face is (010). Crystals are black, highly lustrous, and very hard. Thin sections are colorless in strong transmitted light. The symmetry is orthorhombic; unit cell dimensions and space group are presented in Table I. The nonintegral number of formula weights per unit cell is attributed to statistically distributed vacancies [4]. The crystal structure of this phase, as well as those of the remaining higher borides of aluminum, is not yet known, since preliminary crystallographic studies in the system have thus far been concerned with phase characterization and symmetry relationships.

(c) α-AlB_{12} — This is the most common higher boride phase. It was referred to in the early literature as "graphite-like boron" [5,6]. The symmetry had been described as monoclinic [7] and orthorhombic [5,8], but the true tetragonal, pseudocubic symmetry was given by Halla and Weil in 1939 [9]. Single crystals up to 3 mm in largest dimension have been obtained at this laboratory. Crystals occur as hematite-like plates, with surfaces which are somewhat matte relative to those of AlB_{10}. The plate face is (111) of the pseudooctahedron, or more correctly, (101) on tetragonal axes. Twinning on (101) is fairly common. In strong transmitted light thin sections appear orange-red in color. Cell dimensions and space group are included in Table I. The nonintegral Z-number is accounted for on the same basis as was postulated for AlB_{10}. Here again, a structure determination is not yet available. This is the subject of a study being conducted elsewhere [10].

(d) β-AlB_{12} — The early literature described this phase as "diamond-like or adamantine boron" [11,12]; it has also been referred to as a ternary compound of aluminum, boron, and carbon [5]. The true stoichiometry was given by Náray-Szabó in 1936 [8]. The latter author, however, described the symmetry as tetragonal and presented a tetragonal space group; the true symmetry has been found to be

Fig. 1. Pyramidal [010] and tabular (010) AlB_{10} crystals (20X). (After Kohn, Katz, and Giardini [4]. Reprinted by courtesy of Academic Press, Inc., N.Y.)

Fig. 2. Electron micrograph of polysynthetic twinning as seen on (101)-(011) of β-AlB_{12} (32,500X).

orthorhombic, pseudotetragonal [4] (cf. data in Table I). Single crystals, up to 2 mm in largest dimension, commonly show a bipyramidal habit ($\{101\}$ and $\{011\}$), although doubly terminated octagonal prisms have also been observed. Crystals are generally translucent amber, with the color ranging from yellow to brown. Particularly characteristic of β-AlB_{12} is the ever-present polysynthetic twinning on (110) and (1$\bar{1}$0) [4]. Every crystal of this phase examined thus far has been polysynthetically twinned. The twin lamellae are extremely fine (approx. 2000 A) and sufficiently regular to act as a grating for visible light, resulting in continuous spectra [13]. The high degree of regularity of the twin lamellae is shown in the electron micrograph reproduced in Fig. 2. The mechanism giving rise to this effect is complex and comprises the subject of a separate study. Structure determinations, it should be borne in mind, must first cope with the twinning problem.

(e) γ-AlB_{12}—This phase represents the most recent addition to the aluminum boride family of structures, having been discovered at this laboratory in 1959. In color and general appearance, it is similar to α-AlB_{12}. The symmetry is orthorhombic. Crystals range up to 2 mm and occur as laths or plates. In either case, the better developed face is commonly (100), although (001) has also been noted as a plate face. The relationship of alpha to gamma is one of a polytypic nature. The new phase was first observed on single-crystal x-ray diffraction patterns of α-AlB_{12} in syntactic intergrowth with the more common phase. Attention had initially been drawn to these particular crystals by anomalous morphological data. Figure 3 shows the relationship of the alpha and gamma unit cells. The solid lines outline alpha-phase unit cells projected on (100). Dashed lines designate an alternate choice of axes for α-AlB_{12}, in this case giving rise to an orthorhombic cell with dimensions $a = 17.5$ A, $b = 24.8$ A, $c = 10.2$ A. The gamma phase has its orthorhombic axes along the dashed lines of Fig. 3. Unit translation distance along the 24.8 A direction, however, extends through only two lattice layers, rather than three. The result is an orthorhombic cell having the dimensions $a = \frac{2}{3}(24.8) = 16.6$ A, $b = 17.5$ A, $c = 10.2$ A. Geometrical transformation from the alpha unit cell to that of γ-AlB_{12} is accomplished by alternate stacking of (101)α layers; thus the polytypic relationship between the two structures. Gamma-phase crystals are most commonly observed in syntactic intergrowth with α-AlB_{12} crystals. The volume of the gamma unit cell is twice that of the alpha phase; hence the Z-number of 29.

Still another higher boride of aluminum is given in the literature—so-called "monoclinic AlB_{12}" [9]. This phase is in error, however, its unit cell having been shown to be that of the high-tem-

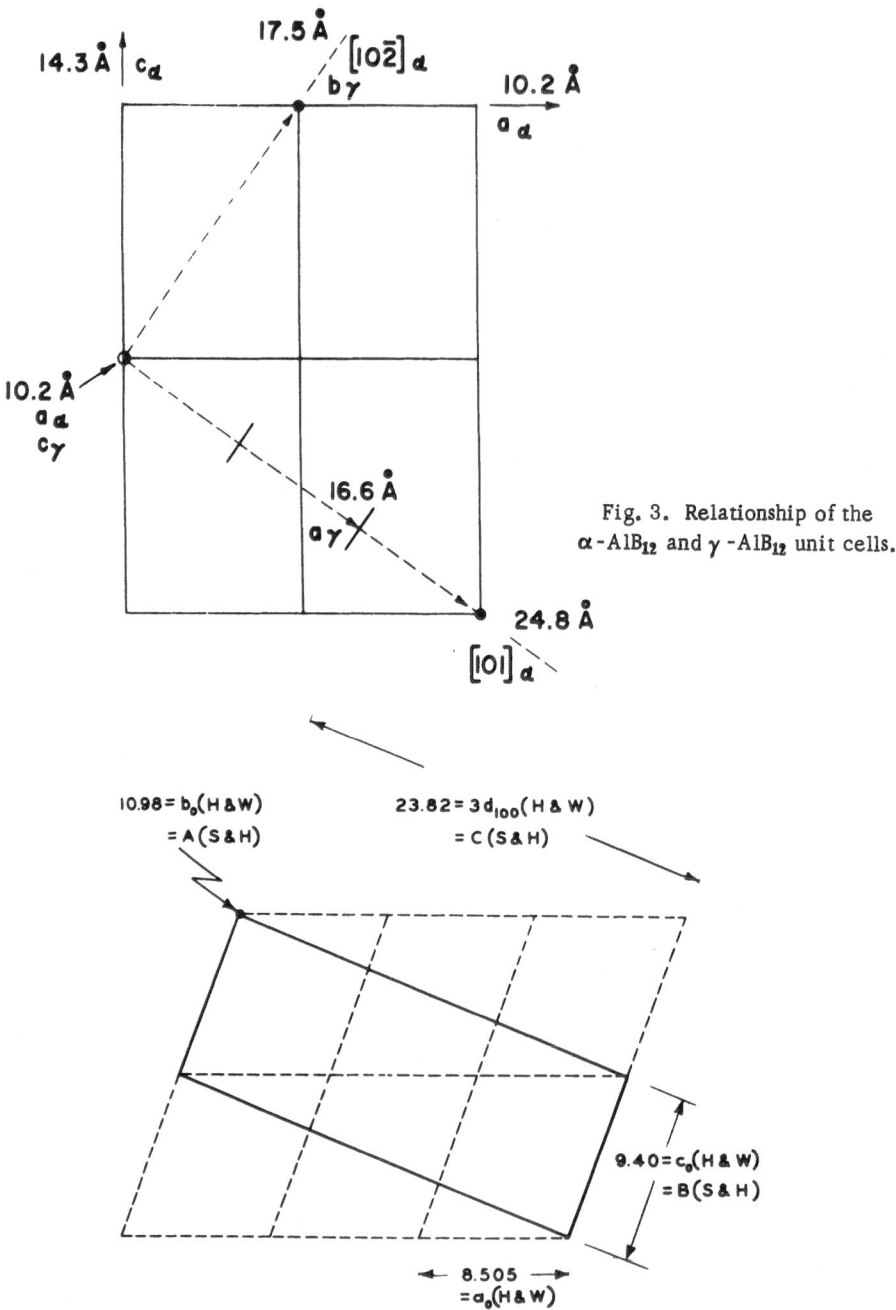

Fig. 3. Relationship of the α-AlB$_{12}$ and γ-AlB$_{12}$ unit cells.

Fig. 4. Relationship of the "monoclinic AlB$_{12}$" cell (9) [(010) projection; dashed lines] to the triple hexagonal cell of beta-rhombohedral boron (14) [(11.0) projection; solid lines]; dimensions are in angstroms. (After Kohn, Katz, and Giardini. Reprinted by courtesy of Academic Press, Inc., N.Y.).

perature (beta-rhombohedral) [14] modification of elemental boron referred to different axes [4,15]. The relationship is shown in Fig. 4.

Boron—Boride Relationships

The boron icosahedron, consisting of a grouping of 12 boron atoms, has long been felt to play an important role in the crystal structure of some of the phases of elemental boron and in some of the higher boride modifications. The structure of boron carbide, for example, has been reported to consist of boron icosahedra alternating with linear chains of three carbon atoms, giving the stoichiometry $B_{12}C_3$ [16]. Alpha-rhombohedral (low-temperature) boron is made up of a close packing of boron icosahedra, with 12 atoms per unit cell [17]. The boron icosahedron has a total diameter of 5.1 A. It is intriguing that a dimension equal to twice this value is common to a large number of phases:

(a) a of α -AlB_{12}
(b) c of β -AlB_{12}
(c) c of γ -AlB_{12}
(d) $2a$ of alpha-rhombohedral boron [17]
(e) $2c$ of tetragonal boron [18,19]
(f) a of beta-rhombohedral boron [14]
(g) a of tetragonal boron [this phase is different from (e)] [20]
(h) approx. $2a$ of boron carbide [16]

It seems almost superfluous to say that a common structural thread underlies the phases listed above. The three known structures of AlB_{12} have other simply related crystallographic vectors, again pointing to common basic structural elements. Whether this extends to AlB_{10} is not yet known.

Summary

The higher borides of aluminum present a series of compounds having complex and highly interesting crystallography. Although it is apparent that these phases bear close structural relationships among themselves and to the modifications of elemental boron, crystal structure determinations have not yet been forthcoming. Practical potential, nevertheless, is considered sufficiently promising to warrant continued study of the structures and structural relationships in the aluminum—boron system.

Acknowledgment

Appreciation is expressed to Dr. A. A. Giardini for the crystals used in the investigation, to Mr. D. W. Eckart, who assisted in the γ -AlB_{12} study, and to Mr. C. F. Cook, Jr., for the electron micrograph.

References

1. Kohn, J. A., Gaulé, G. K., and Giardini, A. A., Proceedings of the Second Army Science Conference (classified), West Point, N.Y., June, 1959.
2. Giardini, A. A., Kohn, J. A., Toman, L., and Eckart, D. W., These Proceedings, p. 140.
3. Hofmann, W. and Jäniche, W., Naturwissenschaften 23 (1935) 851; Z. physik. Chem. B31 (1936) 214-222.
4. Kohn, J. A., Katz, Gerald, and Giardini, A. A., Z. Kristallogr. 111 (1958) 53-62.
5. Biltz, H., Ber. dtsch. chem. Ges. 41 (1908) 2634-2645; Ber. dtsch. chem. Ges. 43 (1910) 297-306.
6. Wöhler, F., Justus Liebigs Ann. Chem. 141 (1867) 268-270.
7. Hampe, W., Justus Liebigs Ann. Chem. 183 (1876) 75-101.
8. Náray-Szabó, St. v., Z. Kristallogr. 94 (1936) 367-374.
9. Halla, F., and Weil, R., Z. Kristallogr. 101 (1939) 435-450.
10. Eriks, K., and Yannoni, N. F., Boston University, Boston, Mass. (NFY now at AFCRC), Private communication.
11. Sella, Q., Ann. Physik 100 (1857) 646-650.
12. Waltershausen, W. Sartorius v., Abhandl. Ges Wiss. Göttingen, Math.-physik. Kl.7 (1857) 297-328.
13. Kohn, J. A., and Eckart, D. W., Anal. Chem. 32 (1960) 296-298.
14. Sands, D. E., and Hoard, J. L., J. Amer. Chem. Soc. 79 (1957) 5582-5583.
15. Parthé, E., and Norton, J. T., Z. Kristallogr. 110 (1958) 167-168.
16. Zhdanov, G. S., and Sevast'yanov, N. G., Compt. rend. acad. sci. URSS 32 (1941) 432-434.
17. McCarty, L. V., Newkirk, A. E., Horn, F. H., Decker, B. F., and Kasper, J.S., J. Am. Chem. Soc. 80 (1958) 2592.
18. Hughes, R. E., Doctoral Dissertation, Cornell University (1953).
19. Sands, D. L., Doctoral Dissertation, Cornell University (1955).
20. Talley, C. P., Post, B., and LaPlace, S., These Proceedings, p. 83.

A NEW MODIFICATION OF ELEMENTAL BORON

C. P. Talley[*], B. Post[†], and S. LaPlaca[†]

A new modification of elementary boron has been prepared by the decomposition of purified BBr_3 on hot W and Re filaments at various temperatures close to 1200°C.

The unit cell is tetragonal; cell dimensions are: $a =$ = 10.12 A, c = 14.14 A. The measured density is 2.364 ±0.005 g/cm^3, indicating that the unit cell contains 192 boron atoms, possibly arranged at the vertices of 16 icosahedra.

Recent work with this phase indicates that the modification is metastable; when melted and subsequently cooled the tetragonal form was found to transform to the rhombohedral modification recently reported by Sands and Hoard. [1]

The polymorphism of boron has been discussed in detail by Hoard [2]. It is generally agreed that three relatively stable modifications exist, i.e., those that have been studied by single crystal methods [1,3,4]. Many other forms of boron have been reported; these are generally considered to be either metastable forms of the element or borides of variable metal content [5].

In the present paper we shall describe a new addition to the list of apparently metastable polymorphs of boron. It should be emphasized that its metastability, per se, should not lead to a lessened interest in its properties or structure. In any event, since the polymorph we are describing appears to be stable up to well above 1400°C, its metastability is of little practical significance.

Boron specimens were prepared[‡] by the hydrogen reduction of boron tribromide on incandescent tungsten and rhenium filaments. The initial filament diameter was 0.025 mm and the final diameter of the boron deposit was about 1 mm. The deposition was carried out at atmospheric pressure and at a temperature of about 1270°C. Contamination by stopcock greases was prevented by using apparatus of Teflon and Pyrex-glass construction without stopcock greases.

Wet chemical analysis of several similarly prepared samples for total boron indicated a boron content exceeding 99% by weight. The

---

[*] Experiment Incorporated, Richmond, Virginia.

[†] Polytechnic Institute of Brooklyn, Brooklyn, New York.

[‡] The specimens were prepared by one of us (C.T.) at Experiment Inc., Richmond, Va. X-ray studies were carried out at the Polytechnic Institute of Brooklyn.

main impurity in the boron rods came from the 0.025 mm diameter metal core and amounted to about 0.7% by weight. From emission spectrographic analysis, aluminum was estimated at 0.005% or less. Traces of other impurities (Ba, Sr, Ca, Cu, Fe, Mg, and Si) were also found which amounted to a total of about 0.02%. The filament method is generally known to produce boron of high purity [6]. The boron rods appeared black when viewed by reflected light, but thin sections appeared red when viewed by transmitted light. Red or reddish crystals of the simple rhombohedral (alpha) form have been observed [4].

The density of the boron was measured by a flotation method and found to be 2.364 ± 0.005 g/cm^3 at 23°C. Density measurements on rod segments and powdered samples indicated that the effect of rhenium or tungsten, if any, on the above value was less than the estimated experimental uncertainty. The unit cell of this material is tetragonal, pseudocubic, with a =10.12 A and c =14.14 A, both ± 0.02 A. The dimensions of this unit cell are similar to those which have been reported for a tetragonal (alpha) modification of AlB_{12} [7].

Table I

d (A)	I/I_1	hkl	d (A)	I/I_1	hkl
8.23	4	101	2.846	3	223
7.15$_5$	8	110	2.806	3	320
6.37	30	111	2.746	35	303/221
5.80	5	102	2.638	18	313/115
5.04	100	112/200	2.608	10	322
4.76	3	201	2.516	10	224
4.52$_5$	5	210	2.456	15	410
4.30	80	211	2.442	25	304
4.114	85	202	2.411	18	411/324
3.935	30	113	2.400	16	215
3.81	45	212	2.386	9	330
3.46	3	221	2.380	9	402
3.358	35	104	2.354	5	331
3.266	14	213	2.263	2	420/332
3.191	3	222	2.233	10	421/401
3.171	3	114			
3.125	2	311	Additional lines not listed		
3.044	4	302			
2.925	5	312			
2.901	4	204			

However, since spectrographic analyses indicate that only trace amounts of aluminum are present in the tetragonal boron, it is evident that the two phases are distinct in spite of the dimensional similarity of the unit cells. When specimens of tetragonal boron were melted by resistance heating and subsequently cooled, they were found to have transformed to the rhombohedral (beta) modification reported by Sands and Hoard [1].

Powder diffraction data are listed in Table I; "d" spacings were computed from patterns recorded on a diffractometer using filtered Co and Cu radiations. Scanning speeds of ⅛ and ¼° (2θ) per minute were used in conjunction with fine (0.003 in.) receiving slits, in order to maximize resolution of closely spaced lines. The pseudocubic character of the material was clearly revealed by the splitting or broadening of a number of lines.

The measured density of 2.364 g/cm^3 indicates that the tetragonal unit cell contains approximately 192 atoms (calc. 190.6), possibly grouped in 16 icosahedra. Efforts are being made to grow single crystals for x-ray diffraction study.

References

1. Sands, D. E. and Hoard, J. L., J. Am. Chem. Soc. 79 (1957) 5582.
2. Hoard, J. L., These Proceedings p. 1; Structure and polymorphism in elemental boron, in From Borax to Boranes (Advances in Chemistry Series) (Washington, Am. Chem. Soc., 1960).
3. Hoard, J. L., Hughes, R. E., and Sands, D. E., J. Am. Chem. Soc. 80 (1958) 4507.
4. McCarty, L. V., Kasper, J. S., Horn, F. H., Decker, B. F., and Newkirk, A. E., J. Am. Chem. Soc. 80 (1958) 2592.
5. a) Náray-Szabó, St. v. and Tobias, C. W., J. Am. Chem. Soc. 71 (1949) 1882.
   b) Lagrenaudie, J., J. Chim. Phys. 50 (1953) 629; Chem. Abs. 48 (1954) 4903.
   c) Rollier, M. A., Proc. XI Congr. Pure Applied Chem. 5 (1947) 935, publ. 1953; Chem. Abs. 47 (1953) 11941 h.
   d) Laubengayer, A. W., Hurd, D. T., Newkirk, A. E., and Hoard, J. L., J. Am. Chem. Soc. 65 (1943) 1924.
6. Powell, C. F., Vapor Plating, Chapter 5, John Wiley & Sons, Inc. New York, 1955.
7. Halla, F. and Weil, R., Z. Krist. 101 (1939) 435.

SOME NEW RARE EARTH BORIDES

I. Binder[*], S. LaPlaca[†], and B. Post[†]

Reaction of rare earth (and yttrium) metal oxides with boron, in the proper proportion, at elevated temperatures, has yielded a previously unreported series of binary rare earth (and yttrium) borides corresponding to the formula MB_{12}.

The MB_{12} preparation is isostructural with the previously reported UB_{12} and ZrB_{12}. The unit cell is face-centered cubic with four formula weights per unit cell.

Details of lattice constants and structures are discussed.

Binary Metal Dodecaborides

The first compound of this type was described by Bertaut and Blum in 1949 [1]. The material is UB_{12}; the unit cell is cubic, space group $Fm3m$, and the cell contains four formula weights of UB_{12}. Steric considerations plus comparisons of observed and calculated intensities indicated that the only variable positional parameter in the structure, that of a boron atom, is $x = \frac{1}{6}$. This leads to a highly symmetrical structure in which the metal atom can be described as

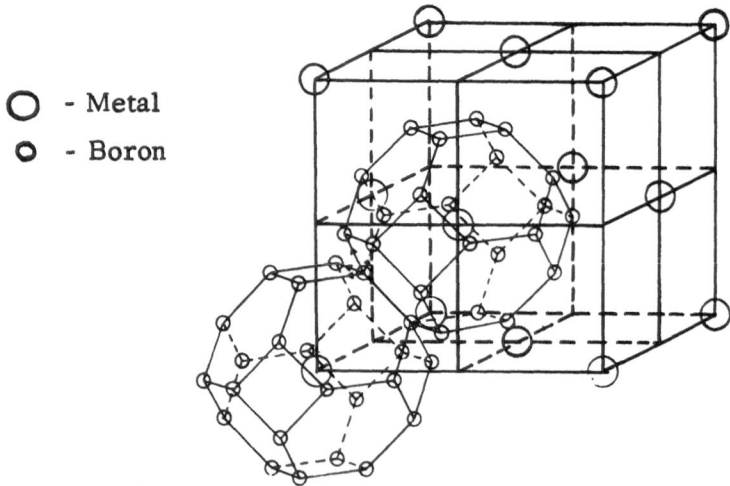

O - Metal

o - Boron

Fig. 1. Arrangement of boron atoms about two adjacent metal atoms in MB_{12}.

[*] Firth Sterling, Incorporated, Yonkers, N.Y.
[†] Polytechnic Institute of Brooklyn, Brooklyn, N.Y.

being in the center of a cuboctahedron having 24 boron atoms at its vertices. Each boron atom is bonded to two metal atoms and to five boron atoms (Fig. 1). The structure may be visualized alternatively (as described by Bertaut and Blum) in terms of a modified NaCl structure, with the metal atoms acting as "Na" and compact groupings of 12 boron atoms in the "Cl" positions.

More recently, Post and Glaser [2] prepared and discussed the isomorphous ZrB_{12}. Good agreement between observed and calculated intensities was computed when the boron atom positional parameter was taken to be ⅙, as in UB_{12}.

Subsequently, Post, Moskowitz, and Glaser [3] described unsuccessful attempts to prepare dodecaborides of the rare earth metals La, Ce, Pr, Sm, Gd, and Yb. In all these cases, no borides with higher boride content than MB_6 were obtained. These were the only rare earth metals (metal oxides) available to the authors at that time.

Recently, other rare earth metal oxides have been made available to us, and efforts to prepare dodecaborides of rare earth and related metals were renewed.

Metal sesquioxides, stated to be 99.95% pure, were heated with amorphous boron in vitrified alumina crucibles at 1400-1500°C in a protective He atmosphere. Reaction times varied from 1/2 to 1 hr. Generally the temperatures were first raised rapidly to avoid formation of rare earth borates (MBO_3). Improved products were obtained when excess boron, sufficient to form MB_{20}, was used and when the oxide plus boron mixtures were ground under acetone prior to the reaction. In this way homogeneous powder mixtures were easily ob-

Table I. Metallic Dodecaborides

MB_{12}	a_0 (A)	$D_{calc.}$	B—B (A)	M—B (A)	R_M^*
LuB_{12}	7.464±0.001	4.868	1.759	2.782	1.902
TmB_{12}	7.476	4.756	1.762	2.786	1.905
ErB_{12}	7.484	4.706	1.764	2.789	1.907
HoB_{12}	7.492	4.655	1.766	2.792	1.909
DyB_{12}	7.501	4.600	1.768	2.796	1.912
YB_{12}	7.500	3.444	1.768	2.796	1.912
UB_{12}[**]	7.473	5.855	1.761	2.785	1.904
ZrB_{12}[***]	7.408	3.611	1.746	2.761	1.888

[*]Implies the "effective radius" of the metal atom in MB_{12} obtained by subtracting $\frac{B-B}{2}$ from M—B.

[**]Bertaut and Blum [1]

[***]Post and Glaser [2]

Table II.   Powder Diffraction Data:
TmB_{12}, CuK_{α}, Ni Filter

I/I_1	d (A)	hkl
100	4.31	111
80	3.74	200
45	2.65	220
100	2.25	311
30	2.16	222
15	1.869	400
30	1.713	331
19	1.670	420
25	1.525	422
11	1.438	511/333
3	1.321	440
20	1.263	531
14	1.246	600/442
8	1.182	620
6	1.140	533
4	1.127	622
2	1.079	444
7	1.047	711/551
5	1.037	640
5	0.999	612
13	0.973	731/553
<1	0.934	800
2	0.9133	733
4	0.9066	820/644
7	0.8810	822/660
4	0.8632	751/555
2	0.8576	662
2	0.8358	840
9	0.8205	911/753
8	0.8157	842
3	0.7970	664
5	0.7836	931

tained; the acetone was evaporated under an infrared lamp. Identification of the reaction products was carried out primarily by x-ray diffraction methods.

The reaction was attempted with oxides of La, Ce, Pr, Nd, Sm, Eu, Gd, Tb, Dy, Ho, Er, Tm, Yb, Lu, and Y; but only Dy, Ho, Er, Tm, Lu, and Y were observed to form dodecaborides. These results confirm the earlier findings of Post, Moskowitz, and Glaser [3].

Dimensions of the cubic unit cells were computed from sharp high-angle reflections using a diffractometer with a scanning speed of $\frac{1}{8}°$ (2θ) per minute. Lattice constants of the six dodecaborides prepared in the present investigation, plus those of UB_{12} and ZrB_{12}, are listed in Table I, together with some additional structural information. Table II lists a typical MB_{12} powder pattern, that of TmB_{12}.

It is evident that size factors play a critical role in the formation of the dodecaborides. The radius of the available hole inside the cuboctahedra formed by the boron atoms ranges from 1.89 to 1.91 A. It has not been found possible up to now to prepare dodecaborides of the larger rare earth metal atoms or of hafnium [4], which is somewhat smaller than zirconium, indicating that the dodecaboride will not form unless the effective size of the metal atom (with a coordination number of 24) lies within, or very close to, the range 1.89 to 1.91 A. Efforts will be made to prepare mixed rare earth metal dodecaborides to test this hypothesis.

Little is known of the properties of these materials. Glaser and Post [5] reported that ZrB_{12} melts at 2680 $\pm$ 100°C and exhibits metallic conductivity (60-100 μohm-cm at 20°C). It appears likely that the other dodecaborides described in this paper have similar properties. These are being investigated.

Measurements of the thermal expansion of erbium dodecaboride from 25 to 1000°C, by use of a diffractometer modified for high temperature, are shown in Fig. 2.

Rhenium Diboride

Rhenium diboride has been prepared by heating the appropriate amounts of metal and boron overnight in an evacuated silica tube at 1200°C (or for 1 hr in an alundum crucible in a He atmosphere at 1400°C). The x-ray diffraction pattern of the product can be indexed on the basis of a hexagonal unit cell with $a = 2.900 \pm 0.001$ A and $c = 7.478 \pm 0.002$ A. There are two formula weights of ReB_2 per unit cell. The x-ray pattern was different from the rhenium boride diffraction patterns published recently by Neshpor, Paderno, and Samsonov [6].

Reflections of the type hhl with l odd were systematically absent. Also, analysis of the intensities of a few reflections indicated that

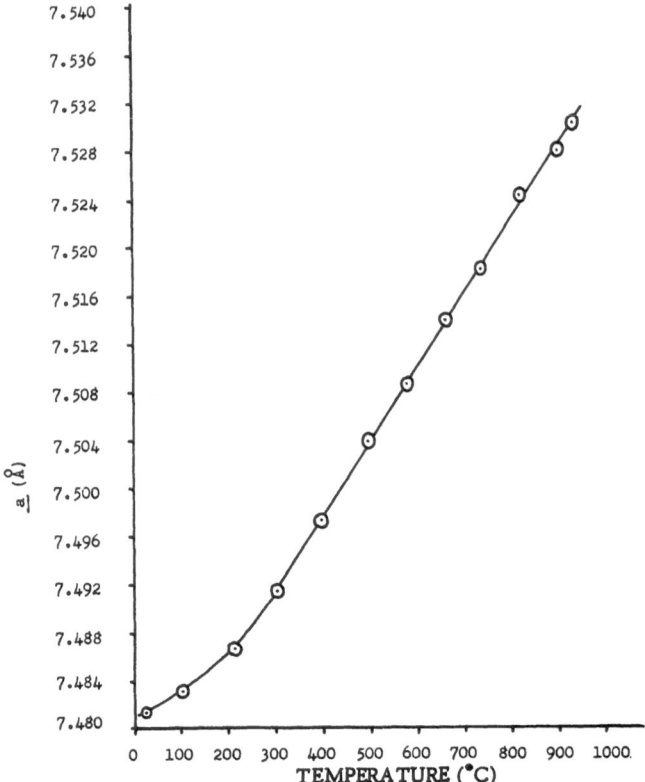

Fig. 2. Thermal expansion of erbium dodecaboride (ErB_{12}).

the rhenium atoms were at ⅓, ⅔, ¼ and ⅔, ⅓, ¾. Steric consider-ations indicated that the boron atoms were in positions at ⅓, ⅔, z; ⅓, ⅔, ½ - z; ⅔, ⅓, $\bar{z}$; and ⅔, ⅓, ½ + z. These positions and systematic absences are characteristic of three hexagonal space groups, but if we assume holohedral symmetry the space group $P6/mmc$ may be assigned to this crystal.

The preliminary analysis outlined above indicates that the ReB_2 structure may be described in terms of relatively close-packed layers of metal atoms separated by layers of boron atoms. The struc-ture differs from the normal AlB_2-type structure in that successive layers of metal atoms do not have the same x and y coordinates, i.e., they do not lie directly above one another. In ReB_2 successive layers of metal atoms (separated by $c/2$) are displaced relative to one another by $a/3$ and $a'/3$. Also, if the z coordinate of the boron atom is ½, then the boron layers will be perfectly planar, as in the AlB_2-type materials. If the z coordinate differs from ½, the boron layers will be puckered. Simple steric considerations place severe limitations on the z coordinates of the boron atoms. B—B bonds in planar nets at $z = ½$ and $z = 0$ would be only 1.676 A long; under these circumstances the Re—B bond length would equal $c/4$ or 1.87 A. Both these bond lengths, and especially the latter, are abnormally small. The effective metallic radius of Re in the elementary metal is close to 1.37 A. In all the AlB_2-type diborides the metal-to-boron distance is somewhat greater than the sums of the radii of the metal and the boron atoms (the latter is generally taken as 0.87 A). This apparently reflects the high coordination number of the metal atoms in this structure; each metal atom is bonded to six other metal atoms plus twelve boron atoms, six in the layer above, and six below. We would therefore expect a Re—B distance of about 2.25 A, rather than the 1.87 A implicit in a structure containing planar boron nets. The 1.676 A B—B separation in these nets would also be far too small; in the AlB_2 structures this distance ranges from 1.74 to 1.81 A [7].

These steric difficulties would vanish if the z coordinate of the boron atom were increased from 0.50 to about 0.55. The shortest Re—B distance would then be 2.245A; the boron nets would be highly puckered, with successive boron atoms within one net at $z = 0.55$ and $z = 0.45$, respectively. The B—B distance would also be increased to 1.835 A.

The analysis of the x-ray data to establish the extent to which the above is true poses serious difficulties. Single crystals were not available. In the powder patterns the intensities of the reflections reflect primarily the overwhelming contributions of the rhenium atoms; the boron contributions are swamped, particularly in the sensitive high-angle region. Useful analysis of the x-ray data was clearly dependent on the precise measurement of the diffraction intensities. Powder patterns were obtained with a diffractometer. Filtered CuK radiation was used. A scintillation counter with pulse amplitude discrimination was used as detector. Patterns were run at ¼° (2θ) per minute. Data are listed in Table III. The areas of the diffraction peaks were measured with a planimeter and these values were converted to F_{obs} in the usual way after subtraction of background.

Structure factors were also computed, after correction of Re scattering factors for dispersion effects, for both $z_B = 0.50$ and $z_B = 0.55$. Comparison with "observed" structure factors indicated that the latter gave better agreement. An ($h0l$) electron density map was then computed. The Re atoms, of course, showed up clearly at the expected position, but there were no unambiguous signs of the boron atoms on this map. Ripples due to the Re atom were evident throughout the electron density map and these were comparable in height, or higher, than the sought-for boron atoms. A difference map was then computed; the Fourier coefficients were $F_{obs} - F_c$, where only the contributions of the Re atoms were included in the calculated structure factors. On this map the boron atoms showed up

Table III.   Powder Diffraction Data:
ReB$_2$, CuK_α , Ni Filter

I/I_1	d (A)	hkl
65	2.74	002
30	2.51	100
100	2.38	101
20	2.08	102
11	1.868	004
40	1.768	103
7	1.499	104
14	1.449	110
15	1.351	112
9	1.284	105
3	1.255	200
1	1.246	006
9	1.238	201
3	1.190	202
9	1.145	114
7	1.121	203
2	1.116	106
2	1.042	204
4	0.983	107
3	0.961	205
2	0.949	210
5	0.945	116
8	0.942	211
1	0.935	008
2	0.9201	212
7	0.8871	213
5	0.8847	206
1	0.8760	108
2	0.8465	214
2	0.8374	300
4	0.8171	302
4	0.8139	207
4	0.8017	215
3	0.7889	109
3	0.7858	118

clearly at $z = 0.54$ and 0.46. Peak heights at the boron positions were considerably higher than residual ripple on this map.

Efforts are being made to grow single crystals to be used in a refinement of the boron parameter. At present the indicated Re$-$B distance is 2.17 A, and the boron-to-boron separation within layers is 1.78 A. The boron layers are highly puckered.

On prolonged exposure to air the diboride undergoes a marked change; the originally black powder changes to one with a striking metallic luster; the diffraction pattern indicates a progressive loss of crystallinity of the ReB$_2$ phase. This effect is being studied further.

References

1. Bertaut, F. and Blum, P., Compt. rend. 229 (1949) 667.
2. Post, B. and Glaser, F. W., Trans. AIME 194 (1952) 631; J. Metals 4 (1952) 631.
3. Post, B., Moskowitz, D., and Glaser, F. W., J. Am. Chem. Soc. 78 (1956) 1800-2.
4. Glaser, F. W., Moskowitz, D., and Post, B., J. Metals 5 (1953) 1119-20.
5. Glaser, F. W., and Post, B., J. Metals 5 (1953) 1117-18.
6. Neshpor, V. S., Paderno, Yu. B., and Samsonov, G. V., Doklady Akad. Nauk SSSR 118 (1958) 515.
7. Post, B., Glaser, F. W., and Moskowitz, D. Acta Met. 2 (1954) 19.

PREPARATION AND PROPERTIES OF MASSIVE AMORPHOUS ELEMENTAL BORON

Claude P. Talley, Lloyd E. Line, Jr., and Quinton D. Overman, Jr.[*]

Massive amorphous elemental boron has been prepared by the reduction of BBr_3 vapor by H_2 in the vicinity of an incandescent tungsten filament 25 μ in diameter. Information on the kinetics of the process was obtained. The deposition apparatus was constructed from Pyrex glass and Teflon fittings in order to prevent contamination from stopcock greases. Conditions were developed for obtaining the boron in the shape of rods up to 2 mm in diameter and 5 to 10 cm in length. Boron deposits in the shape of spheres, hemispheres, and cones were also observed. Wet chemical analysis of a 1-mm-diam. rod for total boron indicated a boron content of approximately 99% by weight.

This type of boron is called amorphous because x-ray diffraction patterns revealed only two diffuse rings. Rods of amorphous boron have shown high tensile strength and Young's modulus ($2.3 \cdot 10^5$ to $3.5 \cdot 10^5$ lb/in.^{2} and $64 \cdot 10^6$ lb/in.^{2}, respectively). The density was determined by a flotation technique and found to be 2.354 ± 0.005 gm/cm^3. This material is very hard, as evidenced by its ability to scratch sapphire.

Amorphous boron exhibits a relatively high resistivity and high negative temperature coefficient of electrical resistance, a characteristic of crystalline boron and semiconductors in general. It also is very opaque in the visible, but can be crystallized by proper heat-treatment into other modifications, including one which transmits a considerable amount of red light.

Introduction

The preparation of elemental boron by "hot-wire" methods dates back many years. The chemical reaction perhaps most often used involves the reduction of boron halides by hydrogen near incandescent filaments and was reported by Weintraub [1] as early as 1911. Through the years a number of investigators have employed this method, using various halides and filament materials of various diameters. However, their primary interest appears to have been

---

[*]Experiment Incorporated, Richmond, Virginia.

the production of crystalline boron. For this reason one gets the impression from the literature that the production of amorphous boron was actually undesirable, and, therefore, it was not actively pursued in the past.

As this paper will demonstrate, amorphous boron has some desirable and unusual properties that merit further study. Our work on amorphous boron thus far has been concerned with its preparation and a somewhat superficial investigation of its properties. To our knowledge, no one has prepared massive amorphous boron as compact cylindrical rods that exhibit the properties we have observed. In the course of preparing boron in its various forms to study its oxidation characteristics, conditions have been found for depositing amorphous boron in the shape of cylinders up to 2 mm in diameter and 5 to 10 cm in length by the reduction of boron tribromide by hydrogen near incandescent tungsten filaments 25 μ in diameter.

Preparation

A schematic diagram of the flow apparatus used is shown in Fig. 1. The apparatus was constructed of Pyrex glass and Teflon fittings to prevent contamination from other materials, such as stopcock greases. As shown in the diagram, H_2 gas enters the apparatus at a given flow rate, temperature, and pressure. It is dispersed through a porous Pyrex glass plug into liquid BBr_3 at a given temperature, picking up BBr_3 vapor, and the mixture passes into the reaction cell. In the vicinity of the incandescent filament the BBr_3 is reduced, boron is deposited on the filament, and the HBr also formed is swept out of the cell by the excess H_2 and any unreacted BBr_3 then condenses in a cold trap for subsequent reuse. A typical set of con-

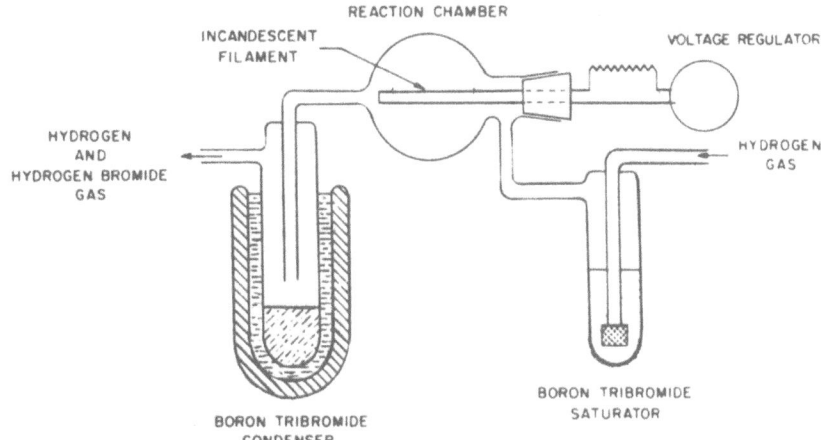

Fig. 1. Schematic diagram of an apparatus for depositing amorphous boron.

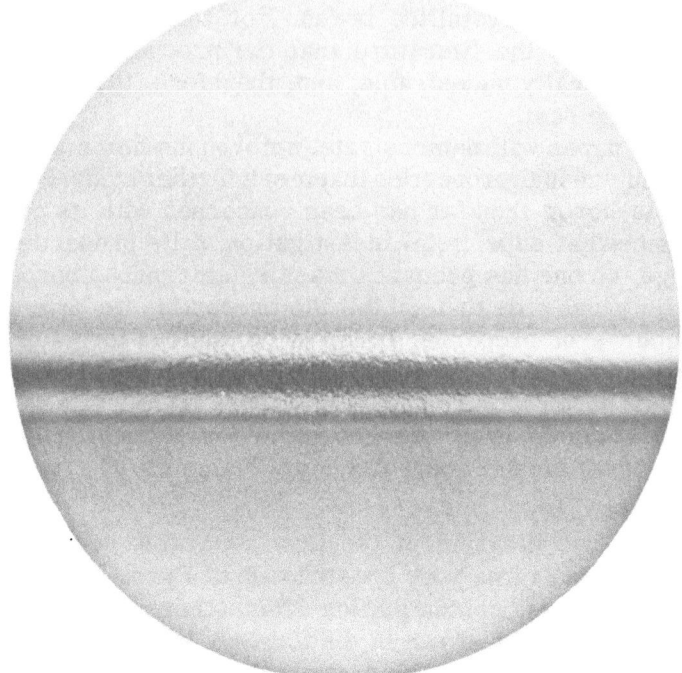

Fig. 2. Amorphous boron rod (approximately 1 mm in diameter).

ditions which gave amorphous boron was: flow rate of 500 cm^3/min at 1 atm and 25° C of H$_2$ previously subjected to O$_2$ and H$_2$O purification; liquid BBr$_3$ temperature of 25° C, which resulted in about 8 to 10 mole percent of BBr$_3$ vapor in H$_2$ downstream of the BBr$_3$ saturator; and a 25-μ-diam. tungsten filament 5 cm in length maintained electrically at 1400°K *. The rate of deposition under these conditions was about 4.5 mg boron/cm^2-min. The rod diameter reached a value of 1 mm in about 25 min. Prior to the deposition, the filament was outgassed in H$_2$ at about 1700°K for 5 to 10 min. A photomicrograph of a rod of amorphous boron is shown in Fig. 2.

The best results from the standpoint of filament breakage have been obtained with tungsten filaments containing 1% ThO$_2$. These filaments presumably do not recrystallize as readily as pure tungsten filaments under the experimental conditions, although we have successfully used other filament materials, such as pure W, Re, Ta, Ti, Mo, and graphite. In addition, the outer surfaces of 1-mm-diam. quartz tubes centrally heated by tungsten filaments have been used as substrates.

---

* Brightness temperatures were measured with an optical pyrometer, and no corrections have been applied because of the lack of accurate emittance data. The true temperatures were approximately 50° K higher than the reported brightness temperatures.

The temperature of the rod during deposition is considered to be the most important factor in the production of amorphous boron. It should not exceed about 1500°K, since otherwise the boron will deposit in polycrystalline form. Below about 1100°K the rate of deposition is relatively slow. These two considerations bracket the desired temperature range for deposition. The logarithm of the rate of deposition as a function of the reciprocal of the filament temperature with 25-μ-diam. tungsten filaments, 500 cm^3/min H$_2$ flow rate, and 25° C BBr$_3$ liquid temperature is plotted in Fig. 3. A temperature dependence corresponding to 15 kcal/mole was calculated from the slope. The interpretation of such apparent activation energies must be made with caution and will not be attempted at this time.

Positioning the filaments horizontally rather than vertically reduces the temperature gradient along the length of the rod and thus aids in the control of the deposition process. Care must also be taken to make the flow of reactants as uniform as possible in order to form a deposit of uniform dimensions.

The flow rate of H$_2$ gas is not particularly critical, and successful deposition has been obtained at flow rates as low as 300 cm^3/min and as high as 1000 cm^3/min. The important factor is that a sufficient mass flow rate of reactants be maintained, so that a reasonable deposition rate can result. This is particularly important when depositing at the higher temperatures so that the total time of deposition is less than that which would allow a solid-state transformation of the amorphous boron to some other form to take place. While solid-state transformation of amorphous boron may be a method

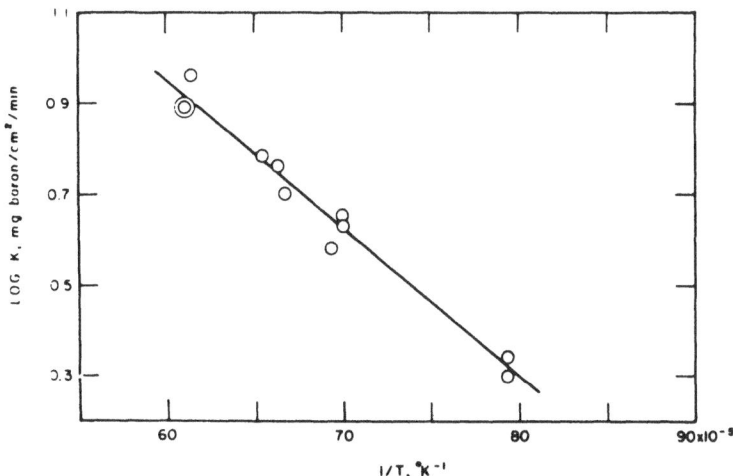

Fig. 3. Logarithm of deposition rate versus 1/T for amorphous boron.

for producing other forms of boron, it is, of course, a deterrent when one wishes to produce amorphous boron.

The mole fraction of BBr_3 vapor in the BBr_3-H_2 gas mixture may not be too critical. While 8 to 10 mole percent has been used in most of our work, concentrations up to the stoichiometric mixture of 40 mole percent and higher should be considered and may possibly give faster deposition rates. It may even be possible to produce amorphous boron rods by the thermal decomposition of BBr_3 alone or by the use of other volatile boron compounds which either decompose thermally or may be reduced at temperatures below 1500°K. The temperature of the BBr_3 is important only for the role it plays in determining the mole fraction of BBr_3 in the BBr_3-H_2 gas mixture.

In addition to rods, deposits of boron in the shape of small spheres, hemispheres, and cones on metal filaments and wrinkles on quartz substrates have also been observed, as shown in Figs. 4 to 7. All of these deposits were formed at 800°K or more below the melting point of elemental boron. At deposition temperatures around 1500°K both polycrystalline and amorphous boron occasionally deposited on different sections of the filament, and at the junction between these, relatively large pieces of material were formed which transmitted appreciable amounts of red light. Density measurements

Fig. 4. Spherical boron growths (approximately 0.1 mm in diameter).

Fig. 5. Hemispherical boron growths (approximately 0.1 mm in diameter).

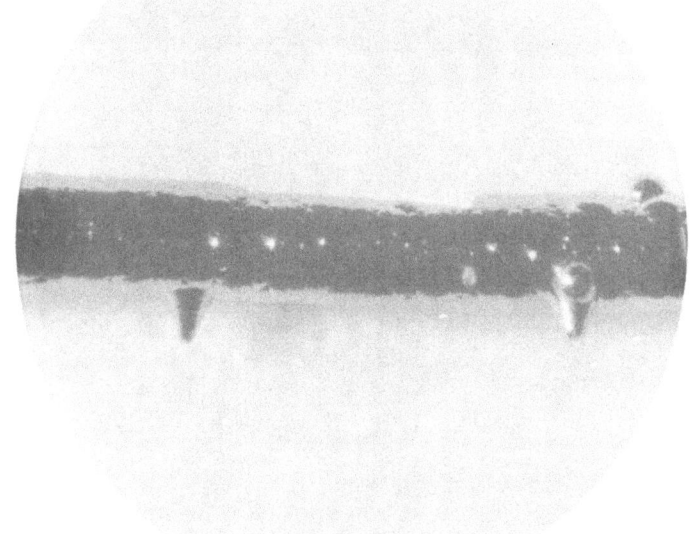

Fig. 6. Conical boron growths (approximately 0.1 mm in height).

indicate that this material is the "low-temperature" (alpha) rhombohedral boron reported by McCarty and co-workers [2].

Properties

Wet chemical analysis of five rods for total boron indicated an average boron content of 98.8% by weight. A large amount of impurity came from the 25-μ-diam. tungsten core, amounting to about 0.5% by weight on the average. X-ray shadowgraphs, visual inspection of cross sections of the rods with an optical microscope, and density measurements of powdered and whole rod segments showed that the tungsten core remained essentially in place, although it may be that trace amounts of tungsten were concentrated in the boron near the filament. Emission-spectrographic analysis indicated in addition to tungsten small amounts of Al, Ca, Fe, Cu, Mg, and Si, amounting to an estimated total of 0.08% by weight, mainly as Si (0.06%).

Rods of amorphous boron have the external appearance of black nails; however, at this point the similarity stops. The density of amorphous boron was found to be 2.350 ± 0.005 g/cm^3 at room temperature, as determined by a simple density-gradient technique. A density gradient was established in a glass tube by uniformly varying the proportions of tetrabromoethane (density 2.97 g/cm^3) and

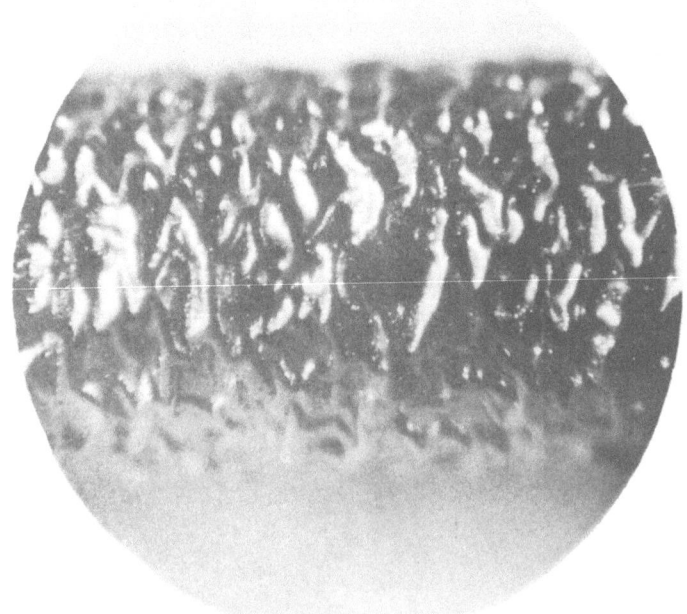

Fig. 7. Wrinkled boron growths (wrinkles are approximately 0.05 mm in width).

chloroform (density $1.49\,\mathrm{g/cm^3}$) from bottom to top and inserting commercially available calibrated glass floats varying in density from 2.3000 to $2.5000\,\mathrm{g/cm^3}$. In this manner a reasonably linear and stable density gradient was established, and the density of unknown boron specimens was determined by noting the depth to which they sank when placed in the tube, as shown in Fig. 8. This method, commonly used to determine density of plastics [3], is quite sensitive, and also has been used to distinguish between some of the various crystalline forms of boron we have prepared.

Amorphous boron is quite opaque to visible light but may be transformed by proper heat-treatment at 1530°K and higher into other modifications, one of which transmits appreciable amounts of red light. Mr. G. K. Gaulé of the U.S. Army Signal Research and Development Laboratory has found that amorphous boron transmits infrared radiation to some extent. X-ray diffraction patterns of

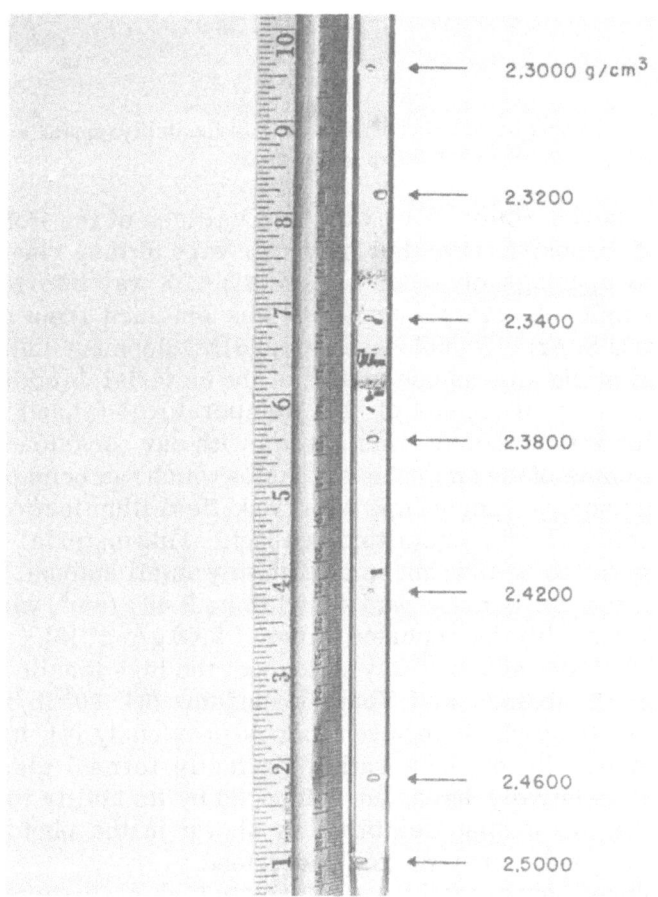

Fig. 8. Density gradient tube containing a number of boron specimens.

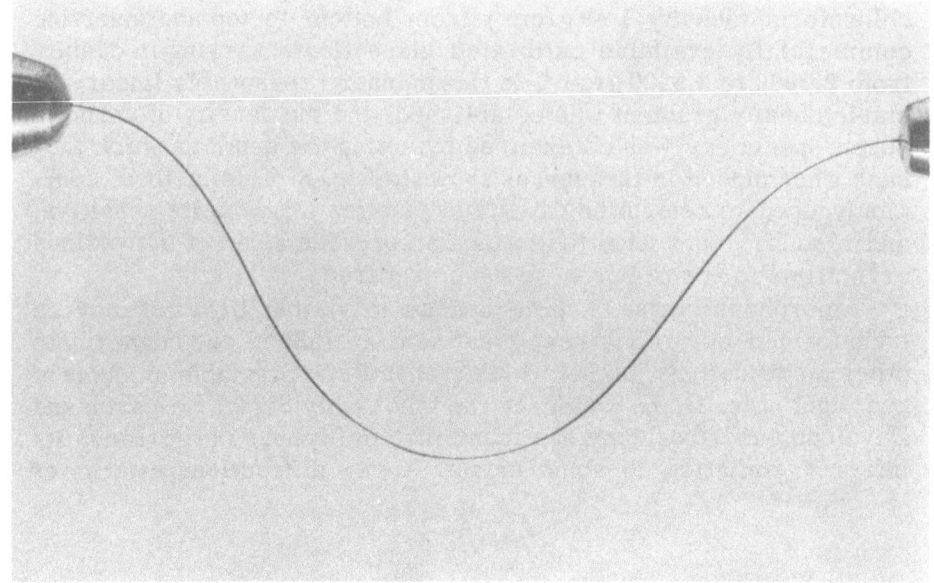

Fig. 9. Amorphous boron rod under load to show flexibility (approximately 0.2 mm in diameter).

amorphous boron obtained by Post and LaPlaca of the Polytechnic Institute of Brooklyn revealed only two very diffuse rings corresponding to spacings of about 2.5 and 4.3 A. X-ray information on another sample of amorphous boron was obtained from Dr. J. A. Kohn of the U.S. Army Signal Research and Development Laboratory, substantiating the amorphous nature of the material; in addition, the presence of a small amount of "low-temperature" (alpha) rhombohedral boron was detected. This agrees with our occasional observation that some of the amorphous samples which had been powdered and viewed under a microscope with dark-field illumination showed a few particles which transmitted red light. This material was visually estimated to be present in a relatively small amount, about 1% by weight, and its density was found to be 2.45 g/cm^3, which is in fair agreement with the reported value of 2.46 g/cm^3 [2].

In view of the above x-ray evidence, the high tensile strength (2.3 to 3.5 · 10^5 lb/in.^{2}) and Young's modulus (64 · 10^6 lb/in.^{2}), and the lack of ductility which we have reported previously [4], it appears that amorphous boron is a rather perfectly formed glass. This material is relatively hard, as evidenced by its ability to scratch sapphire. Its surprising flexibility is shown in the photograph in Fig. 9 of a normally straight rod under load.

Amorphous boron exhibits a relatively high room-temperature electrical resistivity, of the order of 10^4 ohm-cm, and a high neg-

ative temperature coefficient of electrical resistance, a characteristic of crystalline boron and semiconductors in general. A curve
of the logarithm of the resistance versus the reciprocal of absolute
temperature for a small piece of amorphous boron of millimeter
dimensions is shown in Fig. 10.

The thermal conductivity of an amorphous boron rod was determined by a method reported previously for polycrystalline boron
[5]. The tiny 25-μ-diam. tungsten core was heated electrically and
served as the heat source. The circuit consisted of battery, variable
resistor, and the specimen immersed in a constant-temperature water bath. The current through the rod and the voltage drop across the
rod were determined by measuring the voltage drop across precision
resistors with a potentiometer. At the temperatures used in these
experiments the electrical conductivity of the boron was negligible,
and therefore practically all the electrical current flowed through
the tungsten filament located in the core. Electrical contact was
made to the tungsten by fusing on platinum leads at each end of the
rod. Heat generated in the tungsten was conducted through the cylindrical casing of boron and into a surrounding water bath. The tem-

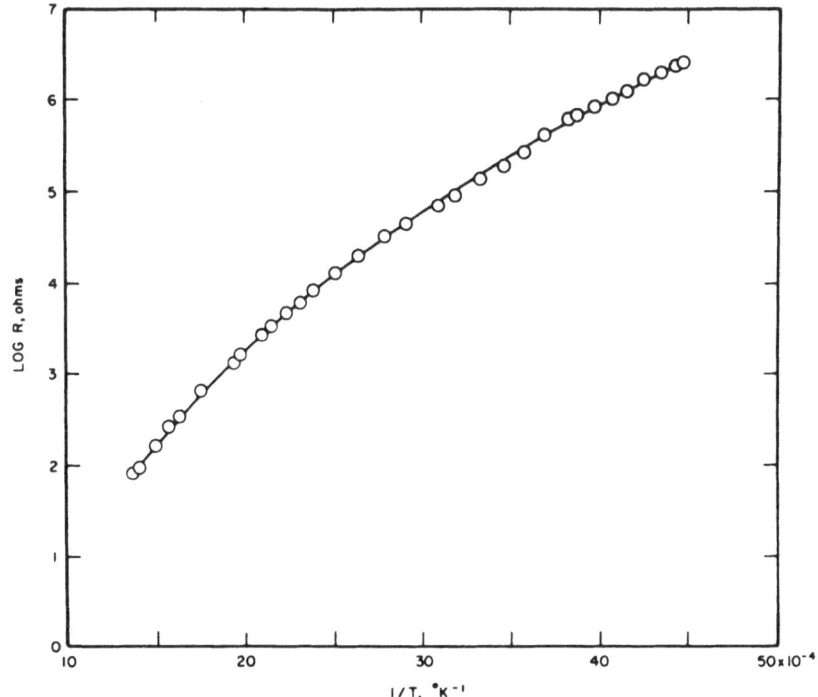

Fig. 10. Logarithm of resistance versus 1/T for a piece of amorphous
boron of millimeter dimensions.

perature of the inside surface of the boron was assumed to be equal to the temperature of the tungsten.

The temperature of the tungsten was obtained by computing its resistance from the measured voltage and current supplied to the rod and comparing the computed value with a curve of resistance versus temperature determined experimentally for the same rod at negligible power input. The temperature of the outside surface of the boron was assumed equal to that of the stirred water bath. In a typical experiment with a 1.86-mm-diam. by 42.0-mm-long rod, the power supplied to the 25-μ-diam. tungsten core was 2.86 cal/sec, the resistance of the tungsten corresponded to a temperature of 92.5°C, and the temperature of the water bath was 31.8°C. Six measurements on this rod at various power inputs gave an average value of 0.00770 cal/(cm^2-sec-deg/cm) with an average deviation from the mean of 0.4%. Considering the accuracy of measurement of the individual quantities involved, the thermal conductivity value obtained is thought to be accurate to within 2%, excluding any systematic errors.

This preliminary work on massive amorphous elemental boron has shown it to possess rather interesting properties. Further investigation of this material should prove equally interesting.

Acknowledgment

We wish to thank the Office of Naval Research for sponsoring part of this work. We would also like to thank Dr. Ben Post and Mr. Sam LaPlaca of the Polytechnic Institute of Brooklyn and Dr. J. A. Kohn of the U.S. Army Signal Research and Development Laboratory for x-ray information on the structure of amorphous boron. We also appreciate the information on the infrared transmission of amorphous boron supplied by Mr. G.K. Gaulé of the U.S. Army Signal Research and Development Laboratory. In addition, we would like to thank Experiment Incorporated and our parent company, Texaco, Inc., for permission to publish this work.

References

1. Weintraub, E., J. Ind. Eng. Chem. 3 (1911) 229-301.
2. McCarty, L. V., Kasper, J.S., Horn, F.H., Decker, B.F., and Newkirk, H.E., J. Am. Chem. Soc. 80 (1958) 2592.
3. "Measurement of Density of Plastics by the Density-Gradient Technique," Am. Soc. Testing Materials ASTM D 1505-57T (1957).
4. Talley, C.P., J. Appl. Phys. 30 (1958) 1114-1115.
5. Talley, C.P., J. Phys. Chem. 63 (1959) 311.

THE FLOATING-ZONE MELTING OF BORON AND THE PROPERTIES OF BORON AND ITS ALLOYS

Earl S. Greiner*

Pressed powder compacts of boron have been melted by the floating-zone technique. The compacts were initially bonded with boric acid, which was at least partially decomposed by heating to 300°C. Subsequently, heating to 600°C completely decomposed the boric acid to boron trioxide, which formed a liquid coating on the particles and after cooling strongly bonded the compact. One end of the compact was fitted into a graphite cup, which was used to heat the adjacent boron to a temperature suitable for inductive coupling within the floating-zone apparatus. A liquid zone was passed from the top to the bottom of the compact in an atmosphere of argon. The boron trioxide evaporated before the temperature reached the melting point of boron, approximately 2000°C.

Thomas and Gutowski [1] have shown metallographically that the zone-melted boron consists of large crystals, many of which are twinned. The structure was that described by Sands and Hoard [2], rhombohedral with 107 or 108 atoms in the unit cell.

The electrical resistivities of two coarsely crystalline specimens of zone-melted boron were identical to those of vapor-deposited boron in the intrinsic conductivity range, but were slightly higher in the extrinsic conductivity range, at corresponding temperatures.

Boron and boron--phosphorus alloys were prepared by the reduction of boron trichloride or boron trichloride and phosphorus trichloride with hydrogen on heated tungsten or tantalum filaments. The impurities in most of the specimens consisted of not more than approximately 0.1% silicon and less than 0.005% of a few other elements.

Phosphorus decreases the electrical resistivity of boron--phosphorus alloys containing 0.0007, 0.080, and 14.7 atom percent phosphorus, respectively, in the extrinsic conductivity range.

The higher phosphorus compound, BP, was prepared by reacting boron trichloride, phosphorus trichloride, and hy-

---

*Bell Telephone Laboratories, Inc., Murray Hill, New Jersey.

drogen at elevated temperatures. The material had the zinc blende structure, and the lattice constant was in good agreement with published values. Williams [3] has prepared BP by other reactions. He reported the forbidden energy gap of BP to be about 5.9 ev.

The high melting point and high reactivity of melted boron present certain difficulties in working with this material in the liquid state. Many of the commonly used refractories are attacked by melted boron. Horn [4] has zone-melted boron in boron nitride boats. Seybolt [5] has prepared boron and boron-rich alloys in this refractory.

It has been shown by several investigators, however, that the floating-zone technique can be used advantageously for melting some refractory materials. Keck and Golay [6], Emeis [7] and Theuerer [8] have used the method for zone melting silicon. Also, Buehler [9] has used the floating-zone method to grow single crystals of tungsten, molybdenum, and columbium. It was believed, therefore, that if suitable compacts could be prepared, the floating-zone technique could be used to melt boron.

Owing to the brittleness and small degree of plasticity of boron at room temperature, the normal methods for preparing pressed powder compacts of boron were found to be unsatisfactory. Instead, compacts of boron powder, suitable for zone melting, have been prepared by using boron trioxide (B_2O_3) as a binder. The powdered crystalline boron (−20 + 100 mesh), obtained from the U.S. Borax and Chemical Corporation, was coated by immersing in a boiling, aqueous solution of boric acid (H_3BO_3) and evaporating to dryness while stirring constantly. The amount of H_3BO_3 was 5% by weight of the boron powder, which gave the compact sufficient strength, after pressing at 100,000 psi, to permit removal from the die without breaking. The compact, $\frac{1}{4} \times \frac{1}{4} \times 8$ in., was then heated in vacuum at 300°C to decompose the H_3BO_3. Subsequently, the compact was heated to 600°C to assure complete decomposition of the H_3BO_3 to B_2O_3 and H_2O. The B_2O_3 formed a liquid coating on the boron particles which, on cooling, solidified and bonded the compact. Later, it was observed that the B_2O_3 evaporated when the sample was heated to the melting point of boron--approximately 2000°C [10].

Preparatory to melting, the compact was mounted in a floating-zone apparatus as described by Theuerer in connection with the floating-zone melting of silicon [8]. In order to zone melt boron, the upper end of the compact was fitted into the mortised end of a piece of high-purity graphite, which was used for heating the adjacent portion of the boron to a temperature suitable for inductive coupling. The melted zone, heated with 3.5 or 4.0 Mc rf current, was passed toward the bottom end of the compact.

In order to prevent the boron oxide which deposits on the inner wall of the quartz envelope surrounding the boron during heating from hindering visual observation of the melt, two concentric transparent quartz tubes as described by Whelan and Wheatley were used [11]. The inner tube contained a slot in its surface adjacent to the boron. Whenever the outer tube became coated with boron oxide, it was rotated until a clean portion of the surface was in line with the slot. A phenol fiber yoke was clamped externally to the top and bottom members of the melting apparatus to prevent rotation of the inner slotted quartz tube during rotation of the outer tube. A water curtain was applied to the outside of the outer quartz tube for cooling. An atmosphere consisting of either 99% by volume of argon and 1% hydrogen or only purified argon was passed through the apparatus during melting. Zone-melted rods of boron up to about six inches in length have been prepared [12].

Thomas and Gutowski have shown metallographically that the zone-melted boron consists of large crystals, many of which are twinned [1]. The structure was that described by Sands and Hoard [2], rhombohedral with 107 or 108 atoms within the unit cell. The electrical resistivity of zone-melted boron will be discussed later.

Boron and boron-phosphorus alloys were prepared by the reduction of boron trichloride or boron trichloride and phosphorus trichloride gases with hydrogen on heated tungsten or tantalum filaments. The amounts of phosphorus in the alloys were controlled by varying the vapor pressure of the liquid PCl_3. The lower phosphorus alloys contained 0.0007, 0.080, and 14.7 atom percent phosphorus, respectively. The higher phosphorus compound BP was prepared by reacting BCl_3, PCl_3, and hydrogen at 900 to 1000°C in a heated Vycor tube. The material had the zinc blende structure and a lattice constant of 4.543 A. Williams prepared BP by other reactions and reported the forbidden energy gap of the compound to be about 5.9 ev [3].

The electrical resistivities of three specimens of vapor deposited boron were determined at temperatures from −160 to 700°C [13]. Spectrochemical qualitative tests revealed that the two purer specimens contained not more than approximately 0.1% silicon and only slight traces or less (<0.005%) of a few other elements. The third specimen, the most impure, contained approximately 2% silicon and traces (<0.03%) of a few other elements. The resistivities of the purer specimens of boron were about 10^{11} ohm-cm at −160°C, 10^6 ohm-cm at room temperature and 0.1 ohm-cm at 700°C. The resistivity of the third and most impure specimen of boron tested was identical to that of the other specimens in the intrinsic conductivity range and differed slightly in the extrinsic conductivity range, at corresponding temperatures. The resistivity of a coarsely crystalline specimen of zone-melted boron was identical to that of the vapor-deposited boron in the intrinsic conductivity range, but was slightly higher than the resistivity of the purer vapor-deposited boron in the extrinsic conductivity range at corresponding temperatures. The energy gap of the vapor-deposited and zone-melted boron was 1.39 ± 0.05 ev at $0°K$. The intrinsic conductivity range begins at about $200°C$.

The electrical resistivities of three boron-phosphorus alloys containing 0.0007, 0.080, and 14.7 atom percent phosphorus, respectively, were determined at temperatures between -160 and $700°C$ [14]. Increasing the amount of phosphorus in boron decreases the resistivity at low temperatures or where extrinsic conductivity predominates. The resistivities of undoped boron and alloys containing 0.0007 and 0.080 atom percent phosphorus, having the structure of boron, are approximately the same at high temperatures. The resistivity of the 14.7 atom percent phosphorus alloy, structurally different from the other materials, is significantly lower at low temperatures and higher at high temperatures than the materials containing less phosphorus. The composition of the 14.7 atom percent phosphorus alloy is $B_{5.8}P$, which is approximately the same as that of the material B_6P, described by Williams [3].

Summary

Pressed powder compacts of crystalline boron, suitable for floating-zone melting, have been prepared by using boron trioxide as a binder. The zone-melted boron was coarsely polycrystalline and some crystals were twinned. The structure was rhombohedral with 107 or 108 atoms per unit cell.

The electrical resistivities of zone-melted boron and vapor-deposited boron were identical in the intrinsic conductivity range and nearly the same in the extrinsic conductivity range. Increasing the amount of phosphorus in boron decreases the resistivity at low temperatures or where extrinsic conductivity predominates.

The compound BP can be prepared by the reduction of BCl_3 and PCl_3 with hydrogen at 900 to $1000°C$.

Acknowledgment

The author is pleased to acknowledge the helpful discussions with Dr. W. C. Ellis and the assistance of Mr. J. A. Gutowski.

References

1. Thomas, E. E., and Gutowski, J. A., Private communication.
2. Sands, D. E., and Hoard, J. L., J. Am. Chem. Soc. 79 (1957) 5582.
3. Williams, F. V., Am. Chem. Soc., Abstracts of papers, Boston Meeting, April 7, 1959, 13M.
4. Horn, F. H., J. Appl. Phys. 30 (1959) 1612.
5. Seybolt, A. U., Trans. A.S.M. 52 (1959) preprint No. 135.

6. Keck, P. H., and Golay, M. J. E., Phys. Rev. 89 (1953) 1297.

7. Emeis, R., Zeit. Naturforsch. 9A (1954) 67.

8. Theuerer, H. C., Trans. Amer. Inst. Min., Met. and Pet. Engrs. 206 (1956) 1316.

9. Buehler, E., Trans. Met. Soc., A.I.M.E. 212 (1958) 694.

10. Cueilleron, J., Compt. rend. 221 (1945) 698.

11. Whelan, J. M., and Wheatley, G. H., J. Phys. Chem. Solids 6 (1958) 169.

12. Greiner, E. S., J. Appl. Phys. 30 (1959) 598.

13. Greiner, E. S., and Gutowski, J. A., J. Appl. Phys. 28 (1957) 1364.

14. Greiner, E. S., and Gutowski, J. A., J. Appl. Phys. 30 (1959) 1842.

SIMPLE RHOMBOHEDRAL BORON—
PREPARATION AND PROPERTIES

F. H. Horn*

An allotrope of boron in a simple rhombohedral crystal structure has been reported [1,2]. Salient features of boron in this
structure are reviewed. Simple rhombohedral boron has been
crystallized from platinum melts. The necessary conditions
are indicated. Some preliminary electrical and optical results for boron in the simple rhombohedral form are discussed and compared with results for boron that has been
melted.

In contrast to the complex rhombohedral crystal structure for
boron that has crystallized from its melt, an allotrope of boron in
a simple rhombohedral crystal structure has been identified [1] and
described in detail [2]. The first preparation involved the pyrolysis
of boron triiodide on a filament kept below a temperature of about
1200°C. The formation of this allotrope from the other boron halides
and from hydrides has since been recognized.

As a semiconductor the simple (alpha) rhombohedral boron is
particularly interesting in two respects: first, it has been obtained
as red crystals--the band gap may be considerably larger than that
for the complex boron or, the simple form may be obtained purer--
and second, the simple form has a structure worked out in detail.
Thus, the electrical properties might be understood in terms of the
structure.

Some investigations have been undertaken using the very small
crystals (¼ mm maximum dimension) obtained from the controlled
pyrolysis of BI_3. These crystals are not always red--we have recovered many that are black, using the procedures referred to. We
have attempted to compare the optical absorption of the red simple
rhombohedral boron with boron in the complex structure. These data
are indicated in Fig. 1. It may be seen that the band edge of the simple form is in excess of 2 ev--representing an increase of at least
several tenths of a volt over the probable band edge for the complex
form as judged from the data for boron of such purity as has been
investigated by us.

The electrical resistivity of the small crystals has been measured and is plotted as a function of reciprocal temperature in Fig. 2.
Electrical contacts were made using platinum. More recent work

---

*General Electric Research Laboratory, Schenectady, New York.

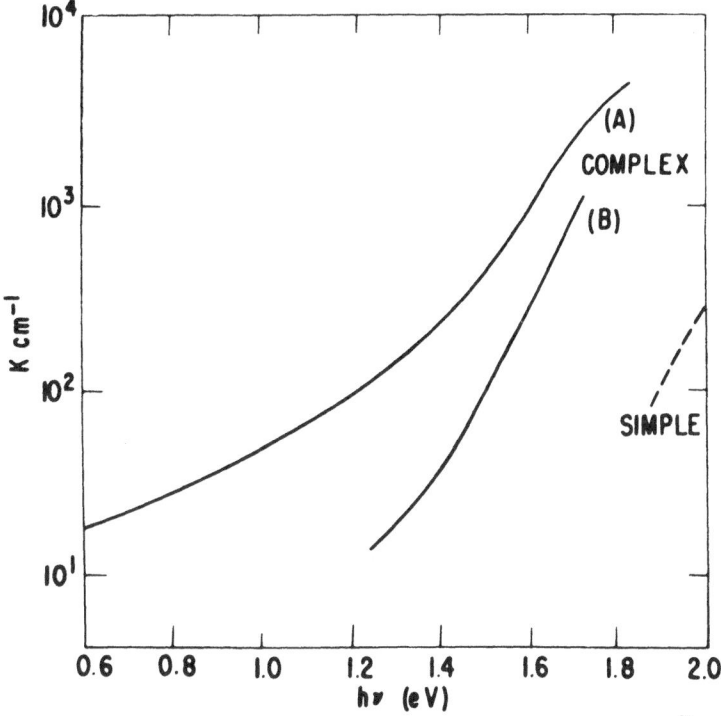

Fig. 1. A and B—complex, rhombohedral boron (B has approx. 10^{18} carriers/cm^3); dashed curve—simple rhombohedral boron.

indicates that at the high temperatures (approximately 750°C) platinum does not make a satisfactory contact. The measurements were
made in hydrogen. The data indicate some results of interest. The
curve of resistivity versus $1/T$ for the red crystals is flat. This
suggests shallow-level impurities and some mobility. Using very
simple arguments based on the optical band gap and the temperature
for intrinsic conduction, one may estimate that the mobility is of the
order of 100 cm^2/v-sec or more. This is greater than has been observed in complex (beta) rhombohedral boron [3]. We also see that
the $1/T$ plot of resistivity for the black, simple rhombohedral boron
(black presumably because of impurities) appears similar to the
data for complex rhombohedral boron. In the case of the simple
form, we may regard the resistivity as being controlled by impurities of the deep-level type. We cannot, of course, say whether the
resistivity of the complex form is clearly due to the same process
since the structure itself may be responsible. However, the two
curves appear similar and the possibility is suggested that the resistivity usually observed for the complex form is controlled by the
presence of deep-level impurities.

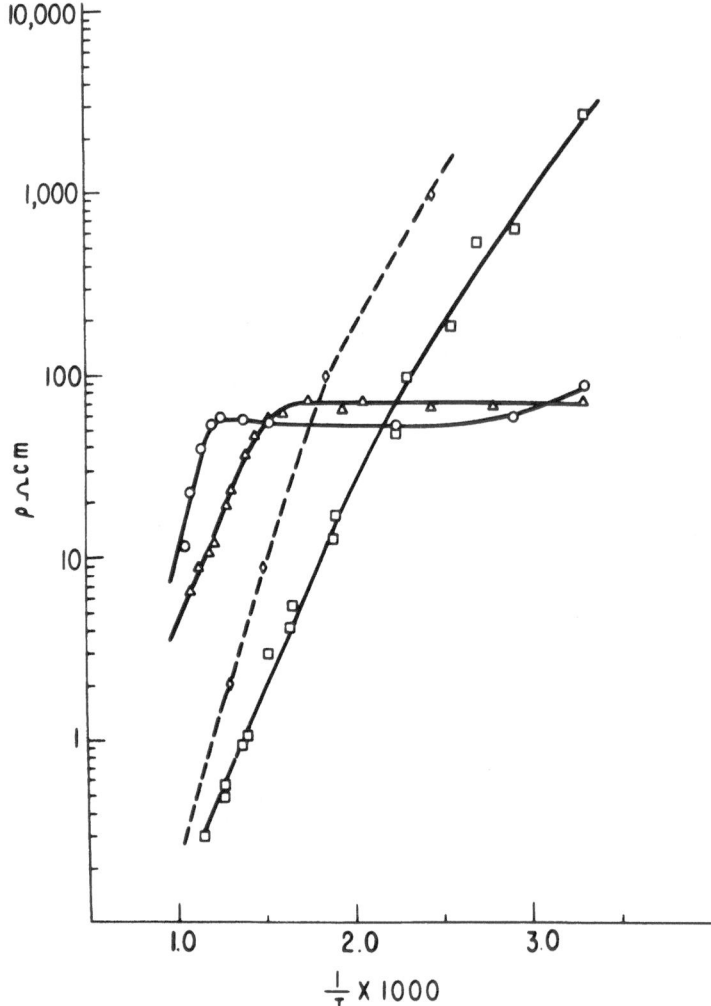

Fig. 2. Circles—crystalline, red boron; triangles—granular, red boron; squares—black, simple rhombohedral boron; diamonds—zone-refined, crystalline, complex rhombohedral boron (approx. 10^{18} carriers/cm^3).

It is known that the simple (alpha) rhombohedral structure for boron converts to the complex (beta) form above about 1300°C. Thus, one approach to investigating the electrical properties of boron is to study the red simple rhombohedral form, convert it to the high-temperature form, and reinvestigate the properties. No large crystals of boron suitable for such an approach have resulted from vapor pyrolysis. Other ways to obtain crystals of the simple rhombohedral form have been sought.

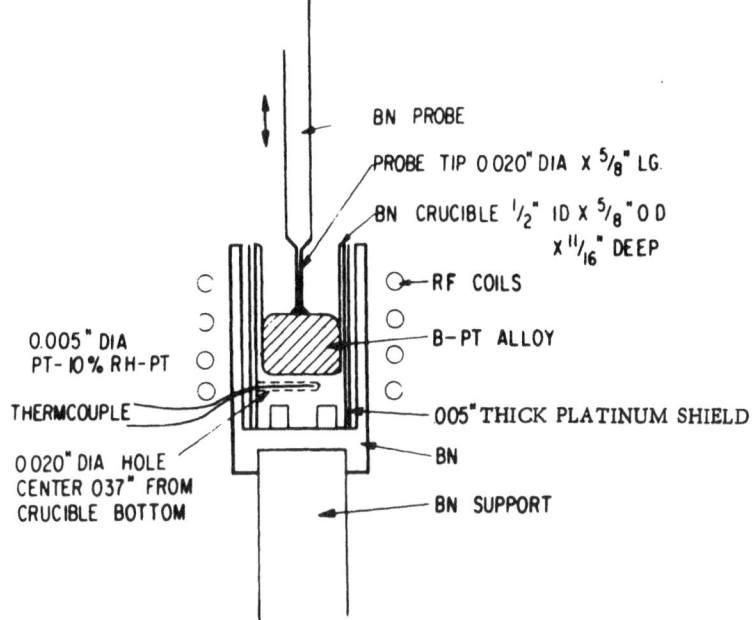

Fig. 3. Apparatus for determining melt temperature of boron—platinum alloys (schematic); system evacuated.

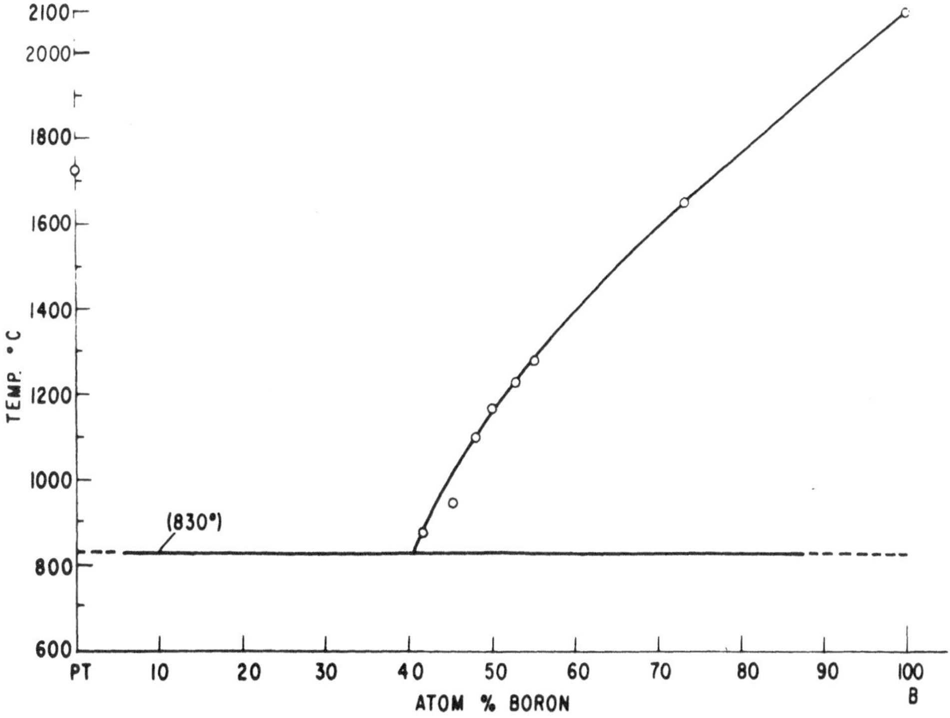

Fig. 4.

In principle, if a solvent for boron could be found from which boron could be crystallized at a temperature at which the simple rhombohedral form is stable, crystals of the simple rhombohedral structure should be recovered.

Boron is known to form a low-melting eutectic with platinum. The phase diagram has not been reported. An attempt was made to determine the phase diagram by the normal thermal arrest procedure. Thermal arrests were observed for the eutectic; but no breaks were detected to indicate the liquidus--solidus. Finally, a probe method was used, the apparatus for which is shown in Fig. 3. For a solution of boron in platinum of known composition, the temperature at which boron would just begin to freeze was noted when the boron nitride probe touched the melt. The temperature at which the probe is withdrawn free of boron was also noted. These temperatures agreed to within about 5°C. Temperatures were measured by an optical pyrometer, and the temperature scale was made to coincide with temperatures measured by a Pt-10Rh thermocouple placed as indicated.

The data for the phase diagram for the boron--platinum system are plotted in Fig. 4. On the platinum-rich side it was not possible to determine the liquidus--solidus because metal probes were attacked and platinum would not adhere to BN or C probes. It is seen from the phase diagram that between the eutectic composition (approximately 40 atom percent boron) and boron there is no indication of compound formation. Thus, from a composition of about 50 atom percent melting at 1200°C, boron should crystallize on solidification of the melt. If the simple (alpha) rhombohedral form is stable, the boron should take on this structure.

Experiments have been conducted to freeze slowly melts of 50 atom percent boron in platinum. Freezing has been allowed to proceed from the bottom of the crucible by lowering the boron nitride crucible (in vacuum) through an induction coil. The lowering rate was 0.04 mm/min. Zone-refined boron and commercial grade platinum were employed. A polished section through such a solidified melt shows black boron at the tip. Dispersed through the platinum matrix but concentrated near the top are crystallites of red boron. Because of the high reflectivity of the platinum matrix these crystals may be missed when illuminated vertically unless the light is correctly polarized. The red crystals have been recovered by crushing the brittle platinum--boron matrix, performing a gravity separation using bromoform, and finally separating the crystals manually under a microscope. The well-formed crystals are very small-- ¼ mm in the largest dimension. X-ray diffraction identifies them as entirely simple (alpha) rhombohedral boron.

Although the simple rhombohedral boron recovered by crystallization is in the form of very small crystals, these experiments suggest several conclusions. First, boron in the simple rhombohedral structure is thermodynamically stable below about 1200°C. Second, a more extended investigation of the crystallization should lead to controlling the crystallization of larger crystals. And third, the crystallization gives red crystals, indicating either that the distribution coefficients for impurities in simple and complex rhombohedral boron differ considerably, or that the conditions for crystallizing the simple form (relatively low temperature) are very favorable to a rejection of impurities by the boron.

Acknowledgment

The author wishes to thank the Journal of the Electrochemical Society and the Journal of Applied Physics for permission to reproduce the figures.

References

1. McCarty, L. V., Kasper, J. S., Horn, F. H., Decker, B. F., and Newkirk, A. E., J. Am. Chem. Soc. 80 (1958) 2592.
2. Decker, B. F. and Kasper, J. S., Acta Cryst. 12 (1959) 503.
3. Hagenlocher, A. K., "Halbleitereigenschaften von Bor," dissertation, Technische Hochschule, Stuttgart (1958).

OBSERVATIONS ON BORON AND SOME BORIDES

W. R. Eubank, L. E. Pruitt, and H. Thurnauer*

Single crystals of beta-rhombohedral boron have been grown in both needle and plate habit from the vapor phase. Maximum size of needle crystals was approximately 5 mm $\times$ 0.1 mm $\times$ 0.1 mm. Penetration twinning and growth steps at one end of a crystal were detected in some cases. These boron needles were characterized by smooth lateral faces of high reflectivity. Thin lathlike red crystals of AlB_{12} were prepared from the elements at reduced pressure.

Dense polycrystalline structures of boron, silicon hexaboride, and strontium hexaboride were prepared by vacuum hot-pressing finely divided powders. Microstructure was studied microscopically with reflected light after diamond polishing and etching.

Change in resistance with temperature was measured for both the mono- and polycrystalline material. Energy gap and intrinsic conductivity determinations from log R vs $1/T$ curves indicate that boron prepared by the vacuum process is of higher purity than commercially available boron and of as high a purity as that prepared experimentally by other workers.

Introduction

Finely divided boron of higher purity than that available commercially was desired as the starting material for preparation of refractory borides. Methods of boron preparation by fused salt electrolysis [1,2] and deposition of boron on a hot substrate by decomposition of a boron halide in the presence of hydrogen [3,4] were considered. The disadvantages of each of these preparative techniques soon became apparent. In the case of fused salt electrolysis, the salts, electrodes, and container were found to contribute to the impurity content of the product. The second method, decomposition of a boron halide on a hot substrate, appeared to offer more promise for a high-purity product. Heated substrates of the refractory metals tungsten, tantalum, and molybdenum, and fused quartz and carbon, even under optimum laboratory conditions, were found to contam-

---

* Central Research Laboratory, Minnesota Mining and Manufacturing Company, St. Paul, 19, Minnesota.

Fig. 1. Needle crystals of beta-rhombohedral boron.

inate the boron to the extent of at least 0.1%. A method of vacuum sintering was found to be a preferred method of preparation. Non-volatile impurity content could be reduced to about 0.1% and the material still remained in a finely divided condition. This had a definite advantage over the halide decomposition method because the extremely hard boron did not have to be further reduced in particle size for use in boride synthesis.

High-purity crystalline boron was prepared by high-temperature, high-vacuum deposition from the vapor phase. The vacuum-purified boron powder was consolidated into hard, dense polycrystalline shapes by a vacuum hot-pressing technique. Properties of both mono- and polycrystalline boron and polycrystalline silicon and strontium hexaborides are considered in the following sections.

Single Crystal Preparation

Crystals of boron were prepared from the finely divided element by vaporization at temperatures of 2030 to 2080°C under a vacuum of 10^{-5} to 10^{-6} mm Hg. The vacuum system employed could handle sizeable outgassing loads at these low pressures, having a pumping capacity of 5000 liters/sec below one micron pressure.

For crystal growth, containers and supports of vacuum-purified boron were fabricated. A cylinder of boron from which tiny needle crystals have grown is shown in Fig. 1. The largest needles are

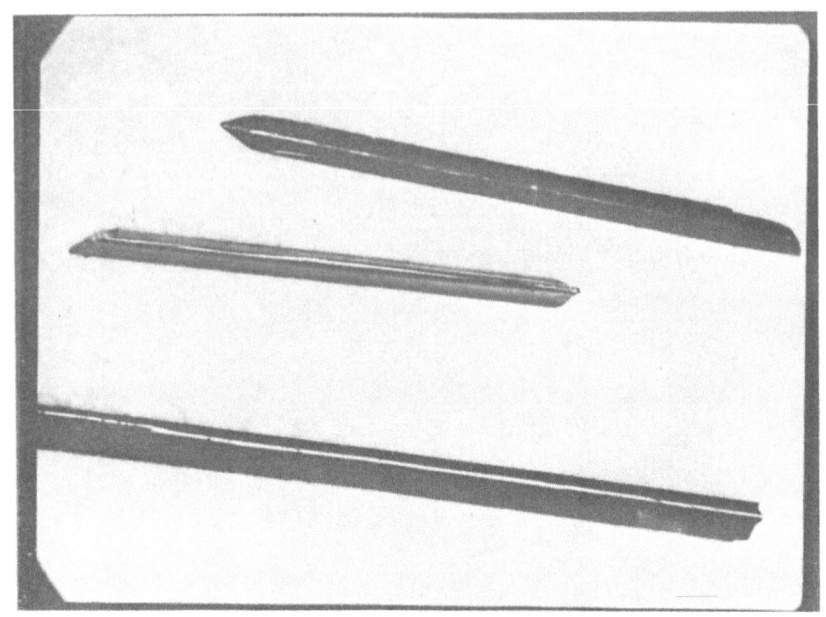

Fig. 2. Boron crystals, 30×.

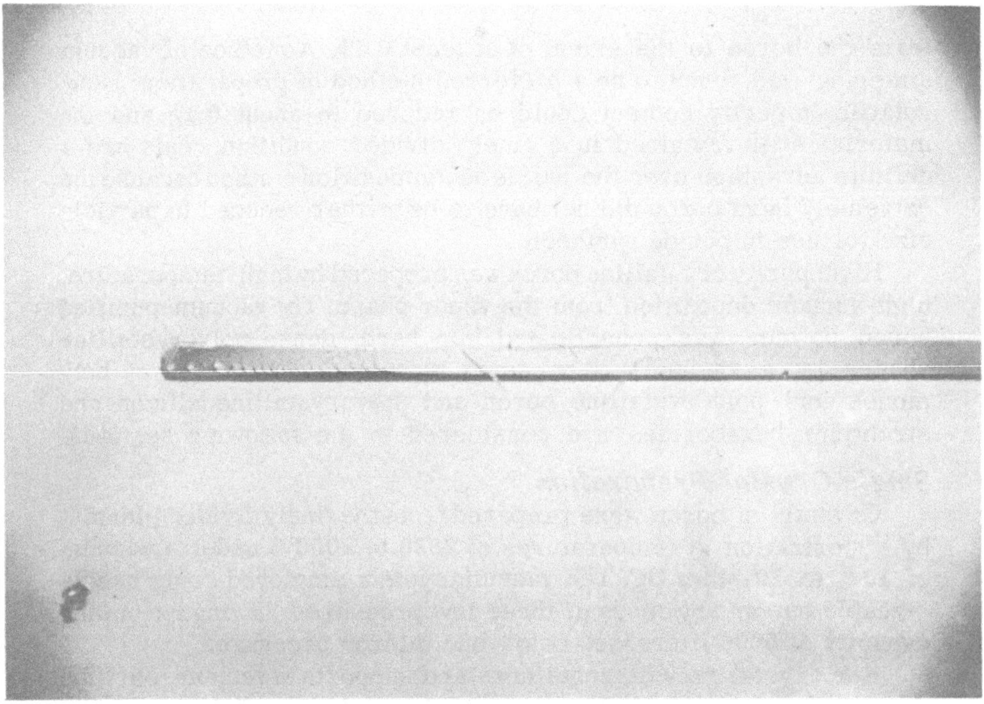

Fig. 3. Growth steps on boron needle, 40×.

Fig. 4. Penetration twinning in boron, 30×.

Fig. 5. Rhombohedral boron platelet, 60×.

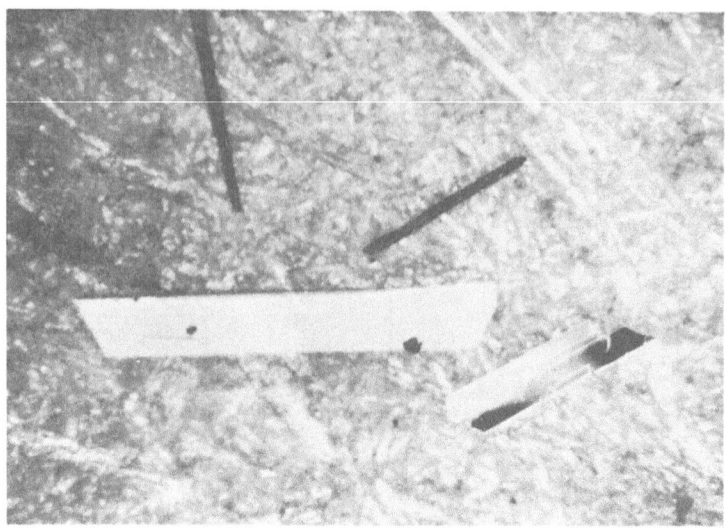

Fig. 6. A lB$_{12}$ crystals, 60

about ½ cm long with a length--width ratio of about 50. A group of crystals broken from the supporting cylinder is shown at a magnification of 30× in Fig. 2. There is considerable variation in size and shape of these needles. Some have smooth crystal faces showing high reflectivity of light. Others appear rounded and circular in cross section. Some of the flatter and larger needles have what appears to be growth steps at one end (Fig. 3).

In rather rare instances penetration twinning, in which one needle grows through another, was observed (Fig. 4). Occasionally tiny rhombohedral platelets were found. Figure 5 shows such a platelet with a circular growth pattern radiating from a defect.

Exceedingly thin lathlike red crystals of AlB$_{12}$ were prepared from the elements (Fig. 6). These crystals were so thin that some light was transmitted. They were very fragile, however, and difficult to handle.

Polycrystal Preparation

Polycrystalline boron was prepared by two methods: vacuum sintering and vacuum hot-pressing of previously vacuum-purified amorphous boron. The vacuum sintering was carried out by heating shapes formed by cold-pressing boron powder with 2½% and 5% B$_2$O$_3$ and water as a binder. These shapes were fired at 1800-2000°C at H$_2$ pressures of 10^{-5} to 10^{-6} mm Hg on boron supports in a tungsten boride-lined graphite container. The resulting structures showed considerable shrinkage and some porosity.

Purified boron powder was vacuum hot-pressed in boron nitride-lined graphite dies at 1800°C at a pressure of about 5000 psi. It was

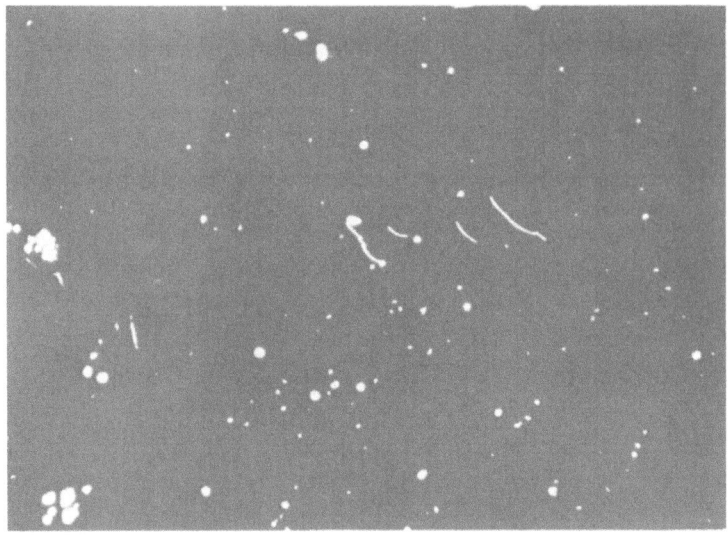

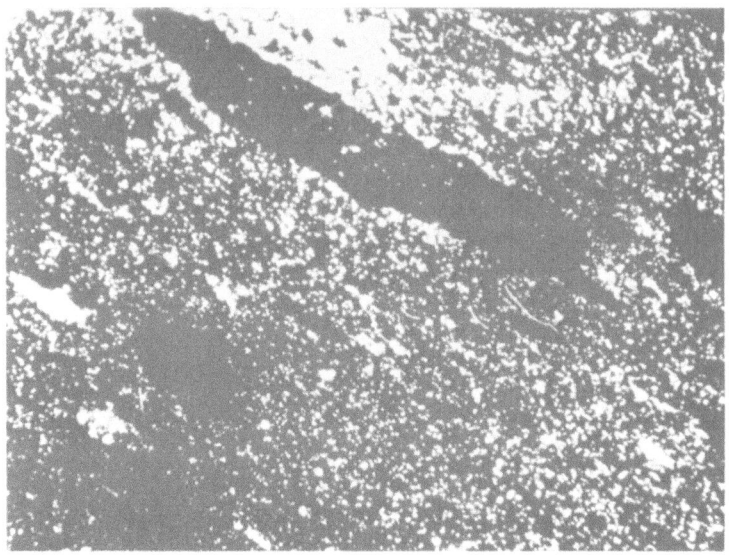

Fig. 7. Vacuum-purified boron, hot-pressed and diamond-polished, reflected light, 200×. a)Unetched field; b) same field etched to show grain structure.

found necessary to outgas these dies to remove B_2O_3 and other volatile materials before use. The hot-pressed boron was higher than 95% of theoretical density. A section of this material was diamond-polished and observed at a magnification of 200× in reflected light (Fig. 7). The surface was quite opaque before etching, showing a few pits from the diamond polishing and a few pores. After etching with 20% KOH−20% $K_3Fe(CN)_6$ solution, some grain structure could be observed (Fig. 7b). A Knoop hardness number (KHN) of 2410 ± 40 under a 100 g load was obtained for this hot-pressed boron.

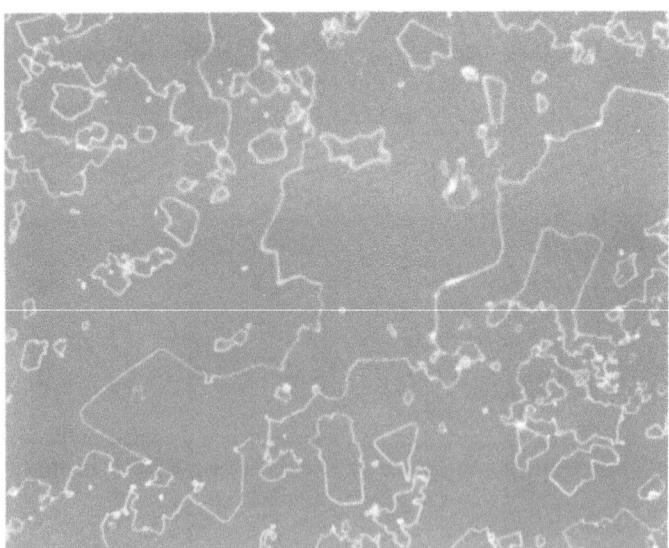

Fig. 8. Hot-pressed SiB_6, diamond-polished, reflected light. a) Etched section with unetched tip, 200×; b) etched section at higher magnification, 575×, showing essentially single-phase grain boundaries.

Silicon hexaboride was prepared by reacting powdered silicon of 99.99% purity with vacuum-purified powdered boron in a boron crucible at 1600°C. The resulting finely divided SiB_6 powder was consolidated by vacuum hot-pressing in a manner similar to that described previously for boron. Observation of diamond-polished surfaces in reflected light (Fig. 8a) revealed a dense, low-porosity structure. Etching revealed a fine grain structure of only one phase. Figure 8a shows an etched section, the tip of which is unetched. A portion of the etched surface at a higher mangification (Fig. 8b) reveals details of the grain structure. A KHN of 1900 ± 40 with a 100 g load was determined for this material. This confirms the findings of Cline [5], who reports a KHN of 1910 for SiB_6. We could not duplicate the work of Samsonov and Latysheva [6], who have reported a KHN of 5352 for a phase they identified as SiB_3.

Strontium hexaboride was prepared by treating freshly calcined SrO with vacuum-purified boron in the presence of carbon at about 2000°C in a boron crucible at reduced pressure. The resulting product was then vacuum hot-pressed at about 1800°C at 5000 psi pressure on the die, which was lined with BN, as described earlier. The resulting pieces of SrB_6 were hard but more porous than the other hot-pressed materials.

Resistance—Temperature Properties

Resistance measurements as a function of temperature were made from room temperature to 1000°K in a small microfurnace described in earlier work [7]. Platinum electrodes were fused to the test samples. Results of the measurements on the various materials studied are given in normalized resistance (log scale) vs one/temperature (absolute) curves shown in Figs. 9, 10, and 11. Vacuum-purified polycrystalline boron, prepared in our laboratory (top curve, Fig. 9) is compared with three commercial polycrystalline borons. The commercial borons were prepared respectively, by decomposition of a boron halide by H_2 on a hot metal filament, on hot carbon, and by arc melting electrolytic boron. The plot for the vacuum-purified boron is much closer to a straight line than the plots for the three commercial borons. The graphs for the commercial borons show considerable curvature from room temperature to about 500°K, indicating appreciable impurity conduction and a narrower range of intrinsic conductivity.

A thermal activation energy of 1.45 ev was obtained for the vacuum-purified boron over the temperature range 500-1000°K. Activation energies for the commercial borons determined over the same temperature range were much lower. Values obtained were 1.27, 1.18, and 0.94 ev., respectively. Of the commercial borons,

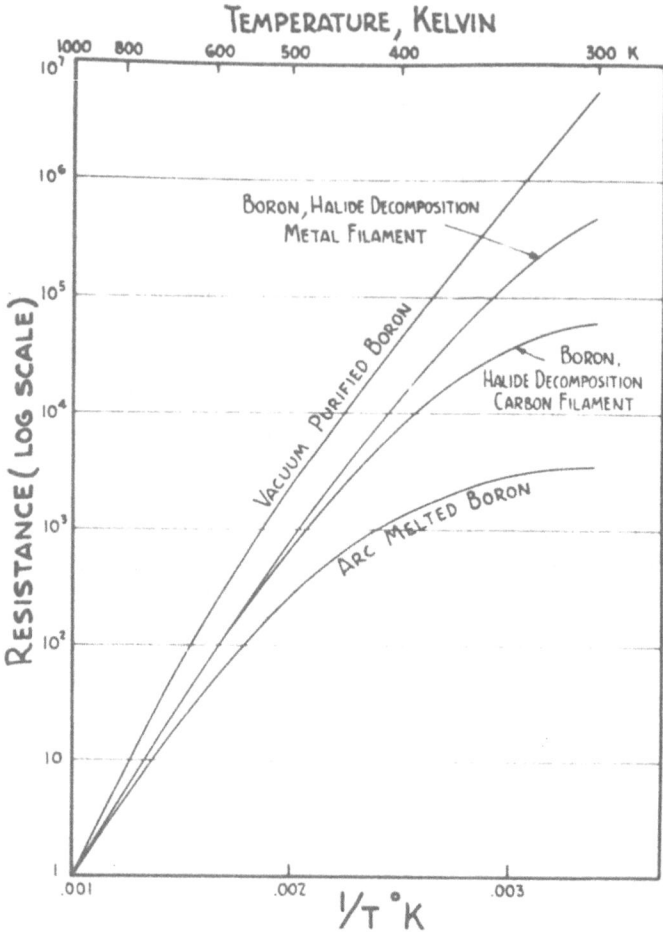

Fig. 9. Logarithm of resistance (normalized) vs $1/T$ (absolute) for polycrystalline borons.

that obtained by decomposition of a halide on a hot metal contained considerably less impurity than that coming in contact with carbon. Boron carbide impurity greatly lowers the resistivity of boron.

Our results for the activation energy agree closely with those obtained by Uno and co-workers [8], who report a thermal energy gap of 1.44 ev over a somewhat narrower temperature range. Greiner and Gutowski [9] obtained a value of 1.39 ev.

Change in resistance (log scale) was plotted against $1/T$ for two needle crystals of boron in Fig. 10. From 500 to 1000°K this shows a straight-line plot, indicating intrinsic conductivity, and the curves for the two crystals are parallel. Below 450°K, impurity levels are indicated by the curvature. A thermal energy gap of 1.23 ev over

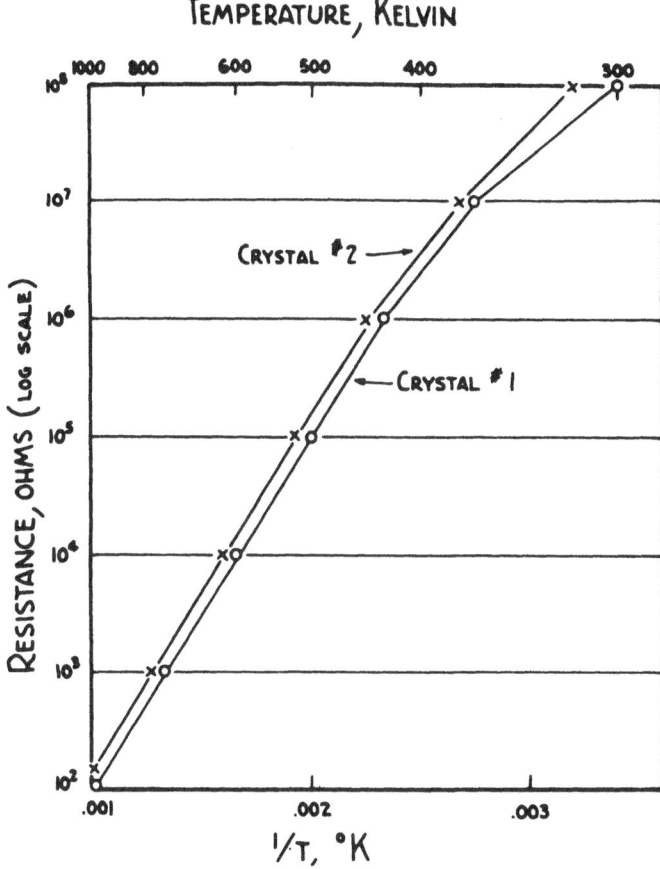

Fig. 10. Logarithm of resistance vs $1/T$ (absolute) for boron needle crystals.

the temperature range 500 to 1000°K was obtained. Shaw and co-workers [10] report a value of 1.55 ev over a much narrower temperature range, 800-1000°K.

A comparison of resistivity change with temperature of strontium hexaboride and a silicon—boron composition is given in Fig. 11. SrB_6 was observed to be a high-resistivity material which undergoes a 10^{-4} decrease over the temperature range studied. Repeated cycles of heating and cooling were found to increase the magnitude of the resistance change with temperature. Four curves indicating four cycles of heating for the same sample are shown in Fig. 11, in which the resistance increased about tenfold between cycles. Repeated heating causes the resistivity to approach that for boron. This is believed to be caused by oxidation of minute amounts of carbon left over from the preparation reaction $SrO + C + 6B = SrB_6 + CO$. The

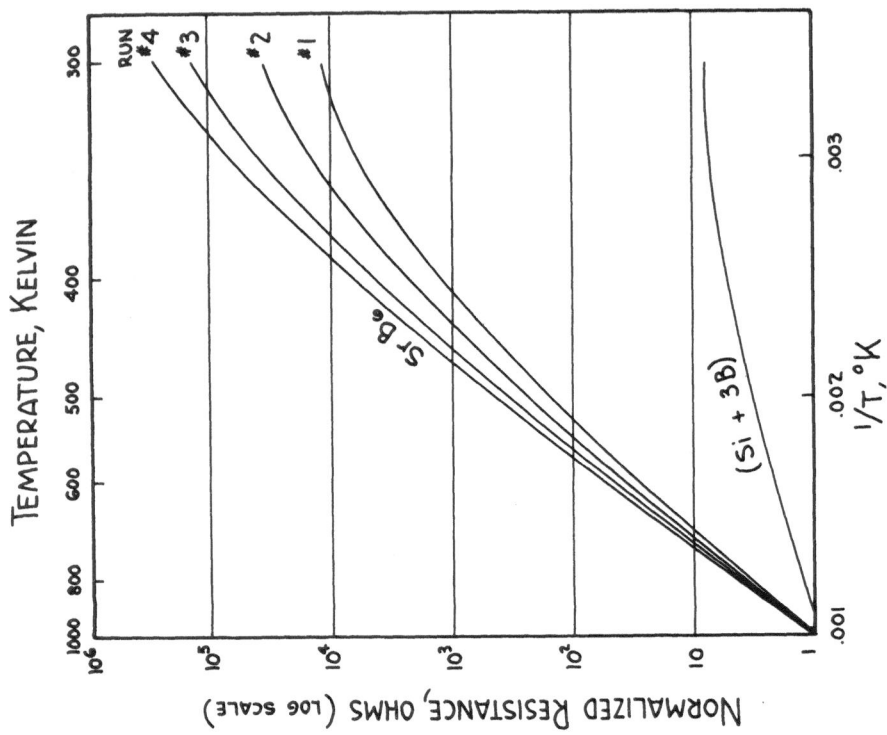

silicon boride, by comparison, is a relatively low-resistivity material, showing much less change with temperature.

Discussion

There are two major difficulties in studying elemental boron. First, there is the difficulty of adequate analytical determination of this element. The best chemical analytical procedures are good to only ±0.4% by weight of boron. High purity is therefore difficult to assess. Indeed, electrical properties appear to be a more reliable indication of boron purity. Based on resistivity determinations of the vacuum-purified boron and spectrographic analysis of the impurity elements present, a boron of 99.9+% purity is indicated. A second difficulty in studying boron is its crystal structure. It exists in at least three crystal forms as well as an amorphous form, and each crystal structure shows a high degree of complexity. Furthermore, because of its low atomic number it has weak scattering power for x-rays.

The needle crystals grown in our laboratory appear to be a different habit of the high-temperature beta-rhombohedral form [11] of boron than has yet been reported. The vacuum-sintered polycrystalline material also appears to be beta-rhombohedral.

The hot-pressing technique in vacuum appears to be a useful technique for consolidating reactive refractory materials. Use of boron nitride-lined carbon dies involves considerable trouble and expense, but the materials involved appear to be the best currently available for this purpose.

Acknowledgment

The authors acknowledge the services of J. C. Close in carrying out many of the experiments. Appreciation is expressed to the Central Research Laboratory, Minnesota Mining and Manufacturing Company, for permission to publish this work.

References

1. Andrieux, J. L., Compt. rend. 185 (1927) 119; Rev. Met. 45 (1948) 49.
2. Cooper, H. S., U.S. Patent 2,572,248 (1951); 2,572,249.
3. Laubengayer, A.W., Hurd, D.T., Newkirk, A.E., and Hoard, J.L., J. Am. Chem. Soc. 65 (1942) 1924.
4. Kiessling, Roland, Acta Chem. Scand. 2 (1948) 707.
5. Cline, C. F., Nature 181 (1958) 476.
6. Samsonov, G. V., and Latysheva, V. P., Doklady Akad. Nauk, SSSR 105 (1955) 455.
7. Eubank, W. R., Pruitt, L. E., and Thurnauer, H., Paper No. 141, 115th Meeting Electrochemical Society (1959).
8. Uno, I. R., Irie, T., Yoshida, S., and Shinshara, K., J. Sci. Research Institute (Tokyo) 47 (1953) 216.
9. Greiner, E. S., and Gutowski, J. A., J. Appl. Phys. 28 (1957) 1364.
10. Shaw, W. C., Hudson, D. E., and Danielson, G. C., Phys. Rev. 107 (1957) 419.
Shaw, W. C., Hudson, D. E., and Danielson, G. C., Rev. Sci. Inst. 26 (1955) 237.
11. Sands, D. E., and Hoard, J. L., J. Am. Chem. Soc. 79 (1957) 5582.

SEMICONDUCTOR PROPERTIES OF BORON[*]

A. K. Hagenlocher[†]

Crystalline boron was grown on a hot tantalum filament in a boron tribromide atmosphere. After removal of the filament, the material was treated by a modified floating-zone process. Silicon is the prevalent impurity in the end product, which was polycrystalline and had large crystallites. The resistivity of purest material was $4 \cdot 10^6$ ohm-cm at room temperature, and $5 \cdot 10^{-2}$ ohm-cm at 1950°C; the intrinsic ionization energy was 1.5 ev. Doped material was also investigated; the ionization energies were 0.6 ev for beryllium, 0.7 ev for carbon, and 0.5 ev for silicon. As expected, doping with beryllium led to p-type, and doping with silicon or carbon, to n-type boron. Low ionization energies found in some highly doped samples suggest the formation of impurity bands when the concentrations exceed approximately $5 \cdot 10^{16}$ cm^{-3}. The Hall mobilities were 55 cm^2/v-sec for holes and 1 cm^2/v-sec for electrons at room temperature. Temperature dependence of mobility is also discussed. Results are compared to those reported by other authors.

The boron used in these investigations was prepared by the thermal decomposition of pure boron tribromide. It was then recrystallized and purified by a method similar to float-zoning. By spectroscopic methods, the purest specimens showed only very small traces of silicon, zinc, aluminum, magnesium, copper, and calcium. Samples were polycrystalline, consisting of crystallites 0.2-2.0 mm in size. Microcrystalline specimens showed the same resistivity as those composed of larger crystallites.

As shown on the accompanying graphs, the conductivity of boron is very low. Based on experiences with silicon, it is possible for a semiconductor to display high resistance owing to a barrier layer between its inversion layer and the inner crystal (Fig. 1). As a result, too small a current density is observed. Measurements were made to determine whether this was the case with boron. A hollow cylinder, serving as one electrode, was placed on the surface of a

---

[*]Extracted and condensed from the author's doctoral dissertation, Technische Hochschule, Stuttgart (1958).

[†]Telefunken, Ulm, Germany.

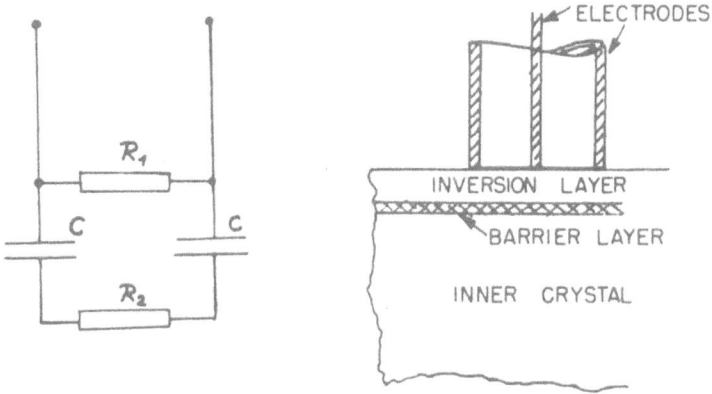

Fig. 1. Electrode arrangement (right) and corresponding circuit (left) for determining the presence of a barrier layer in boron.

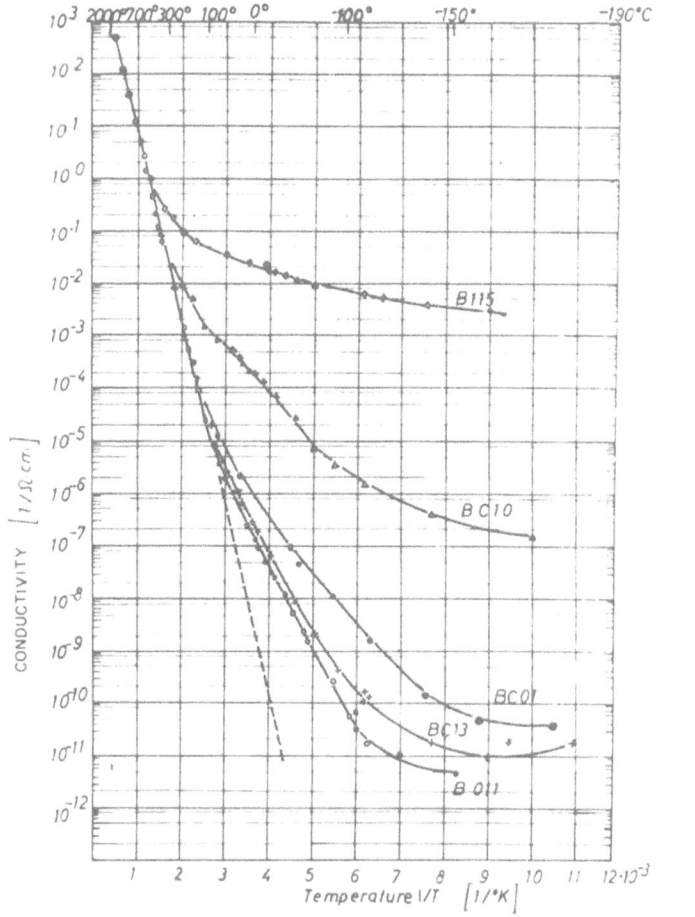

Fig. 2. Electrical conductivity vs temperature for n-type boron.

boron specimen, with the second electrode along the cylinder axis. The parallel reactance and resistance were determined with an impedance meter up to frequencies of 250 Mc. The corresponding circuit is shown in Fig. 1. At higher frequencies the reactance disappears and the real resistance can be measured. The conductivity was found to be the same as that determined by the dc point-contact method. In the latter case, an inversion layer of 1 meg/cm^2 did not affect the measurements if the point-contact pressure was sufficiently high. The dielectric constant appears to be smaller than that reported by most investigators.

Figure 2 presents the temperature—conductivity curves for boron (n-type at room temperature). The lowest curve is for boron which appears to be nearly intrinsic. The other samples are doped with carbon and silicon. The uppermost curve shows the largest amount of impurities, with the middle and lower curves having less

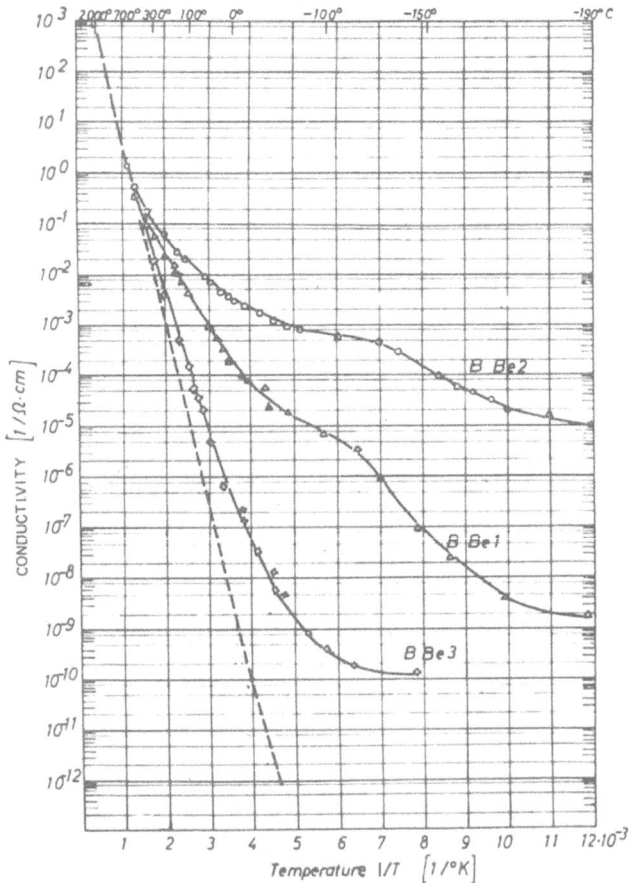

Fig. 3. Electrical conductivity vs temperature for p-type boron.

contamination, in that order. The temperature was varied from −180 to +1950°C, giving a conductivity change over 15 orders of magnitude.

Figure 3 shows the temperature−conductivity data for three samples of Be-doped, p-type (at room temperature) boron. No explanation was found for the slopes of the measured curves.

The Hall constant of n-type boron is plotted against temperature in Fig. 4. With increasing temperature the sign of the Hall coefficient changes and all samples are p-type. The hole mobility exceeds that of electrons. Since the purest sample of boron (B 011) is n-type at room temperature, traces of silicon were suspected; these were found by spectroscopic methods. Be-doped p-type samples (see

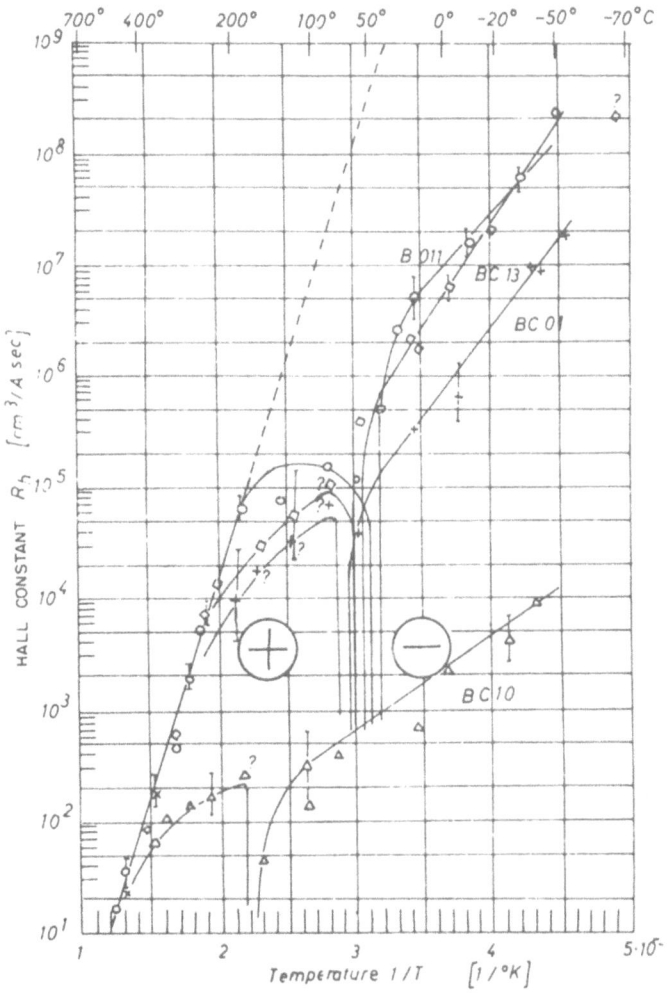

Fig. 4. Hall constant vs temperature for n-type boron.

below) have a mobility nearly 20 times the electron mobility of the purest sample.

For the temperature at which the Hall constant changes sign one obtains

$$n = n_i/b,$$

where n is the electron density, n_i the intrinsic carrier density, and b the ratio of electron and hole mobilities. According to the curves, this condition requires a hole mobility 200 times larger than that of electrons. The electron mobility, however, is 1-1.5 cm²/v-sec and, by extrapolation to 20°C, the hole mobility is 55 cm²/v-sec. Thus, the hole mobility is only 50 times that of electrons, requiring that

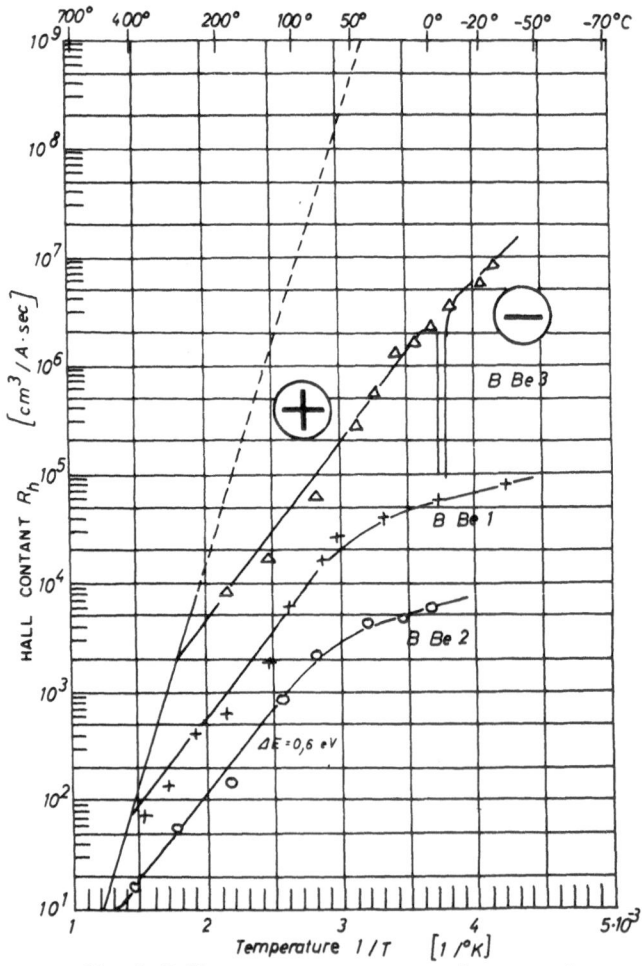

Fig. 5. Hall constant vs temperature for p-type boron.

the Hall constant change sign at too low a temperature. This cannot be explained by assuming acceptors in boron, since such levels would always be filled with electrons produced by donors. Shaw and co-workers [1] also found that the Hall constant sign is required to change at too low a temperature.

Figure 5 presents the Hall constant data for Be-doped p-type boron. Sample B Be 3, having the lowest level of doping, shows a change in the sign of the Hall coefficient at 0°C. Perhaps the more highly doped samples show sign changes at lower temperatures, but it was not possible to make measurements in that range.

An energy gap of 1.6 ev was found from the Hall constant data. The exponent of temperature dependence of hole mobility is approximately ⁷/₄. Impurity levels in boron are as follows:

Carbon (donor)   0.6-0.7 ev below conduction band
Silicon (donor)   0.5 ev below conduction band
Beryllium (acceptor)   0.6 ev above valence band.

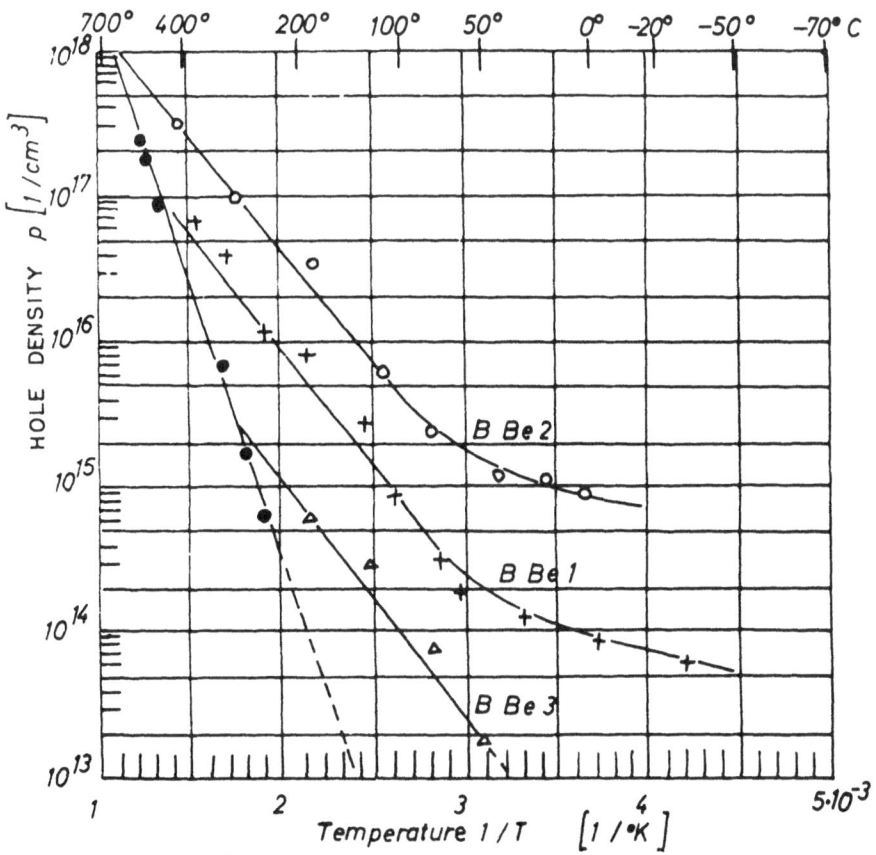

Fig. 6. Hole density vs temperature for p-type boron.

Beryllium may also have a second level 0.2 ev above the valence band. The more highly doped carbon samples show a lower ionization energy; the formation of impurity bands is suggested. Boron shows triboluminescence, indicating the presence of free carriers or dipoles in the lattice; such a case is unknown in semiconductors.

Figure 6 shows the dependence of carrier density on temperature in p-type and intrinsic boron, as determined from the Hall coefficients by the usual relations. The impurity density from these curves is 10^{18} cm^{-3} for sample B Be 1 and $5 \cdot 10^{19}$ cm^{-3} for sample B Be 2.

Assuming an isotropic medium, the energy gap and carrier density of intrinsic boron would be related by the formula

$$n_i = 2\left(\frac{2\pi m_0 kT}{h^2}\right)^{3/2} \left(\frac{m_n m_p}{m_0^2}\right)^{3/4} e^{-\Delta E_0/2kT} \tag{1}$$

if $m_n m_p = 3300 m_0^2$ (m_n and m_p are the effective masses of electrons and holes, respectively; m_0 is the rest mass of an electron). Not only are the required effective masses too large, but their values are not compatible with the impurity densities. The formula for impurity density is

$$N_D = \frac{n^2}{2}\left(\frac{2\pi m_n kT}{h^2}\right)^{-3/2} e^{\Delta E_0/kT} \tag{2}$$

Plausible results are obtained only if m_n and m_p are between 0.1 and 10. In a hydrogen-like model, $m_n \approx 0.1 m_0$. An attempt was made to solve the problem by assuming that there are two pairs of carrier charges, i.e., two kinds of electrons and two kinds of holes, each with a different mobility. Each pair would have its own energy gap, ΔE_{01} and ΔE_{02}, one being assumed larger and the other smaller than the ordinary energy gap, ΔE_0. The pair with the larger gap has a smaller temperature dependence of mobility. The Hall constant and carrier density values would be changed; the carrier density of intrinsic boron is now smaller than that shown in Fig. 6. Equation (1) is satisfied without requiring such large effective masses. However, the use of this model to explain the temperature at which the Hall constant changes sign is possible only if

$$n_1 \cdot p_1 = n_{i1}^2 , \qquad n_2 \cdot p_2 = n_{i2}^2 . \tag{3}$$

In other words, both pairs of carrier charges must always be in balance. Such a case is unknown in the solid state and does not seem too probable. One concludes, then, that it is impossible to explain these boron results on the basis of an ordinary cubic lattice semiconductor model.

Reference

1. Shaw, W. C., Hudson, D. E., and Danielson, G. C., Phys. Rev. 107 (1957) 419-427.

SOME ETCHING STUDIES ON BORON

Ray C. Ellis, Jr.*

Samples of crystalline boron were polished and reacted with various etchants. Several etching compositions are recommended because of the rather unusual reducing ability of boron.

A series of etching studies was undertaken on polycrystalline and single crystal boron. The purpose of the studies was to develop useful etches that would: (1) clean the surface; (2) reveal crystal boundaries, twin planes, and single crystal areas in polycrystalline material; (3) assist in sample orientation; and (4) assist in the study of crystal defects. The polycrystalline boron was purchased from the Pacific Coast Borax Company and single crystal material was grown in the Research Division.

The chemical and physical properties of boron are very similar to those of silicon. Both are hard and brittle, transmit infrared, and are semiconductors. Both form hard carbides, volatile halides, and acidic glass-forming oxides.

Unlike silica, boron oxide is somewhat soluble in water and should not protect elemental boron from attack in aqueous solutions. This is not the case, however. In general, boron is more resistant to aqueous solutions than is silicon. Curiously, boron is not attacked by hydrofluoric acid, by hydrofluoric-nitric acid mixtures, or by boiling sodium hydroxide or phosphoric acid solutions. However, unlike silicon, boron is slowly attacked by boiling sulfuric or nitric acid. Figure 1 shows a sample of polycrystalline boron that has been etched for 10 min at 300°C in sulfuric acid. The crystals, crystal boundaries, and twin planes are clearly developed. Apparently an oxide layer builds up to different thicknesses over differently oriented material to produce colored areas which are seen as the different grays in the photograph.

Figure 2 shows a similar sample etched in boiling nitric acid for 10 min. Voids, polishing scratches, and some scraped surface oxide can be seen. Figure 3 shows the same sample after the same etch had been continued for one hour. The boron was almost randomly attacked, leaving a rough irregular surface.

Several fused salt mixtures were used to etch boron. Figure 4 shows a sample that was etched in a mixture of $1\,NaOH:1\,KNO_3$, in

---

*Research Division, Raytheon Company, Waltham, Mass.

Fig. 1. Boron etched in H_2SO_4 at 300°C for 10 min.

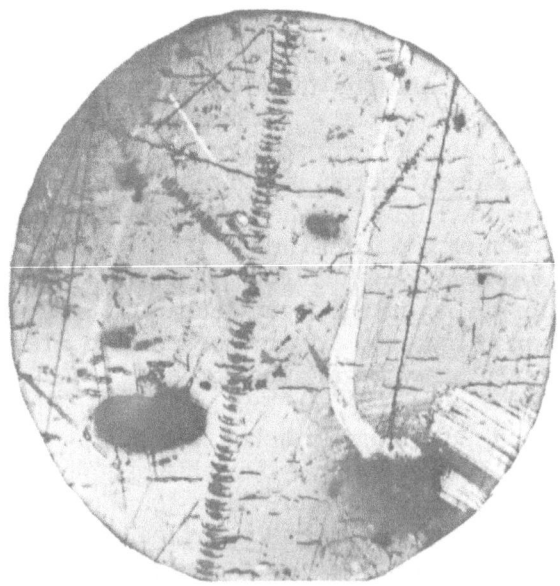

Fig. 2. Boron etched in boiling HNO_3 for 10 min.

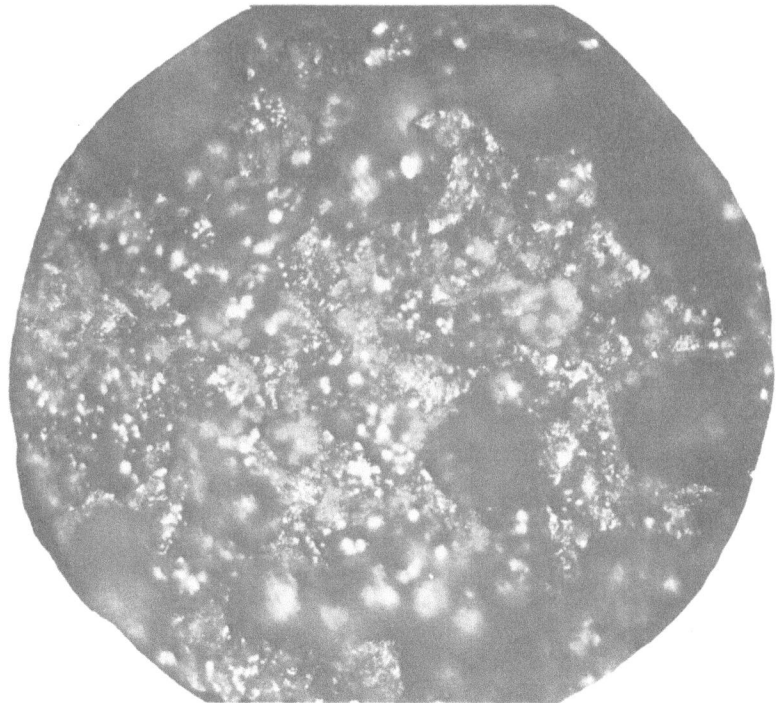

Fig. 3. Boron etched in boiling HNO_3 for 70 min.

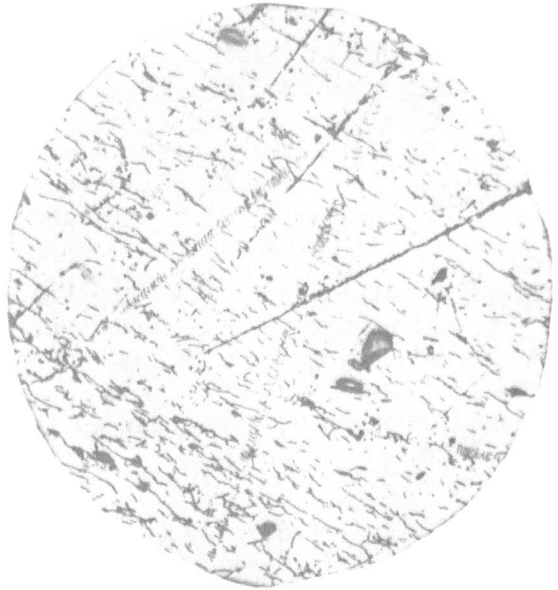

Fig. 4. Boron etched in mixture of 1 NaOH to 1 KNO_3
by weight at 400°C for 10 sec.

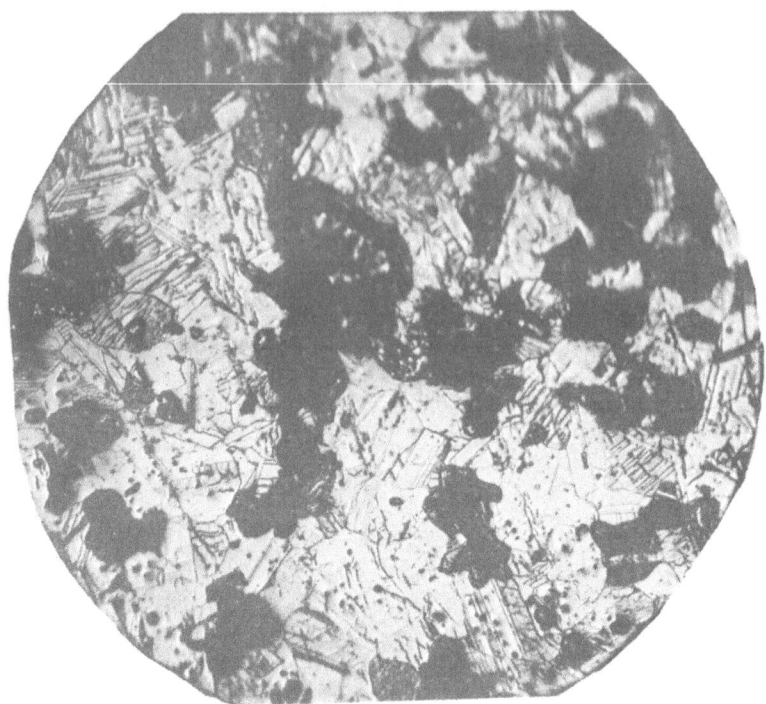

Fig. 5. Boron etched in mixture of $1\,Na_2B_4O_7$ to $1\,KNO_3$ by weight
at 620°C for 1 min.

parts by weight, for 10 sec at 400° C. A series of moon-shaped
cracks in the boron surface are revealed, which were apparently
caused by scratching during polishing. It can be seen, again, that
the attack was most rapid along polishing scratches. Figure 5 shows
a sample of polycrystalline boron that was etched in a mixture of
$1\,Na_2B_4O_7 : 1\,KNO_3$ for 1 min at 620° C. Preferential attack at grain
boundaries, twin planes, and scratches can be seen.

During preliminary studies and compounding of etch compositions
it was found that carbonate-containing etches left a black deposit
of graphite on the surface of boron. Etch compositions containing
sulfate were analyzed after reaction with boron and found to contain
polysufide. The equations representing these reactions are

$$4B + 3Na_2CO_3 \longrightarrow 3C + Na_2B_4O_7 + 2Na_2O$$

$$16B + 7K_2SO_4 \longrightarrow 4K_2B_4O_7 + 3K_2S + 4(S)$$

These reactions take place even in the presence of oxidizing agents
such as KNO_3 or $KClO_3$. Undoubtedly, the reaction is limited by diffusion in the fused salt near the boron surface. Therefore, etch

compositions containing carbonates are not recommended because of the graphite precipitate left on the boron surface.

The unusual reducing property of boron has also been illustrated by the rapid reaction of boron with most fused salt etches. If the temperature is not regulated carefully and the mixture overheats, the boron sample frequently becomes incandescent and is consumed rapidly.

Boron, like silicon, reacts normally towards many etchants, but is inert toward others that, theoretically, are equally reactive. This can be explained at the present time only by saying that the activation energy of these inert reactions is very high.

VECTOR HARDNESS PROPERTIES OF BORON AND ALUMINUM BORIDES

A. A. Giardini, J. A. Kohn, L. Toman, and D. W. Eckart*

The results of a detailed investigation on the vector hardness properties of the aluminum--boron system are reported. The experimental method used was that of Knoop-type micro-indentations. Data are presented for the (10.1) plane of boron, the (110), (101), and (221) planes of α-AlB_{12}, the (100-010),† (101-011),† and (201-021)† planes of β-AlB_{12}, the (100) plane of γ-AlB_{12}, the (010), (101), and (111) planes of AlB_{10}, the (00.1) plane of AlB_2, and also for polycrystalline aluminum and boron. AlB_{10} has been found to possess both the greatest magnitude of microindentation hardness and the greatest anisotropy in the materials studied. A hardness phase diagram, crystallographic and vector azimuth illustrations, and data tables are included.

Introduction

As part of a continuing program at this laboratory directed toward the search for new high-performance electronic, optical, and magnetic materials, a systematic investigation of fundamental properties is being carried out on the various compounds of the aluminum−boron system. At present, seven distinct phases are recognized in this system. They are Al, AlB_2, AlB_{10}, α-AlB_{12}, β-AlB_{12}, γ-AlB_{12}, and B. The latter five possess a high degree of both physical and chemical stability. Their interest with respect to possible application in communication devices has been discussed in an earlier paper [1]. A crystallographic description of the aluminum borides also has been reported [2, 3]. This work represents a part of the above-mentioned basic study of the aluminum−boron system and constitutes the first systematic investigation of the vector hardness properties of these materials.

Data based on Knoop-type microindentations are presented for the (00.1) plane of AlB_2, the (010), (101), and (111) planes of AlB_{10}, the (110), (101), and (221) planes of α-AlB_{12}, the (100)-(010)†, (101)-(011)†, and (201)-(021)† planes of β-AlB_{12}, the (100) plane of γ-AlB_{12}, and the (10.1) plane of boron. In addition, hardness data also are presented for polycrystalline aluminum and boron. Melting temper-

---

*U.S. Army Signal Research and Development Laboratory, Fort Monmouth, N.J.
†Polysynthetic twinning.

atures have been determined and are included to show the observed correlation with hardness for the compounds under study.

Specimen Preparation

The aluminum sample used was of 99.999% purity, polycrystalline, and in rod form. No single-crystal material was available at the time this work was carried out. Single crystals of all the aluminum borides were grown from melts using both commercially pure aluminum (approximately 99% purity) and boron (85-90% purity) and also commercial aluminum with technical-grade B_2O_3. Both alumina and graphite crucibles were used. Temperatures ranging from 1400 to 1700°C were employed. Selected crystals were recovered from the reaction matrices after suitable treatment with relatively concentrated acids. In addition, some of the AlB_{10} crystals investigated were grown by an aluminothermic reaction involving aluminum, B_2O_3, sulfur, and $KClO_3$ in fireclay containers. The single-crystal boron (99.9% purity) was pulled from an induction furnace melt, and the polycrystalline boron (99.9%) was grown by the hot-filament (tantalum) technique.

Polycrystalline specimens were mounted in plastic holders and single-crystal samples were cemented on thick glass disks with Canada balsam. Crystallographic orientation was carried out by back-reflection Laue x-ray diffraction patterns. Limits of orientational error were maintained below 3 degrees of arc. All surfaces were prepared for hardness studies by grinding and polishing through several stages of diamond paste down to 0-3 μ particle size upon cast iron lapping blocks.

Experimental Apparatus

A Bergsman microhardness tester equipped with a certified Knoop diamond point was used to make indentations. The instrument was mounted on a Vickers projection microscope. The latter was set on rubber pads to reduce the effect of building vibrations.

An indenter load of 100 g was used for all measurements. Loading time for indentations was approximately 20 sec. The optical system consisted of a standard filar eyepiece in combination with a 0.25 N.A. 10× objective for aluminum, a 0.65 N.A. 40× objective for AlB_2, and a 1.30 N.A. 100× oil immersion objective for AlB_{10} through boron. Each combination was chosen to produce an indentation image of suitable size and clarity. Filar graduations of the eyepiece were calibrated with a stage micrometer.

Experimental Data

A total of 1110 indentations were used in this study. Only those free of fractures and of perfect apparent symmetry are included. Data for each crystallographic plane were obtained from at least

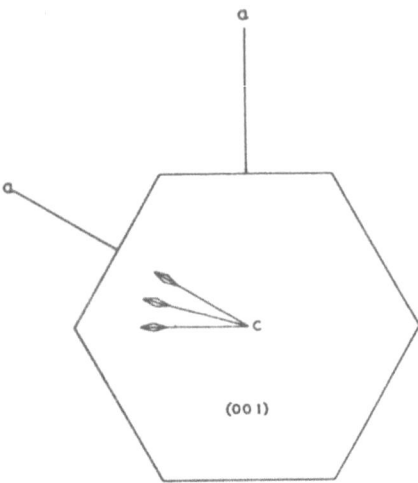

Fig. 1. A crystallographic sketch of the only observed habit of AlB_2. Crystal symmetry is simple hexagonal. Azimuth orientation is defined as parallel to the long diagonal of the indentation.

Fig. 2. Photograph of typical AlB_2 crystals. The latter generally form as very thin multilayered hexagonal flakes having a bronze-like luster. Basal cleavage is very pronounced.

two crystals. In some cases as many as seven crystals of a given phase were studied. Each hardness vector shown is derived from the average of three readings on at least five accepted indentations along the specified azimuth.

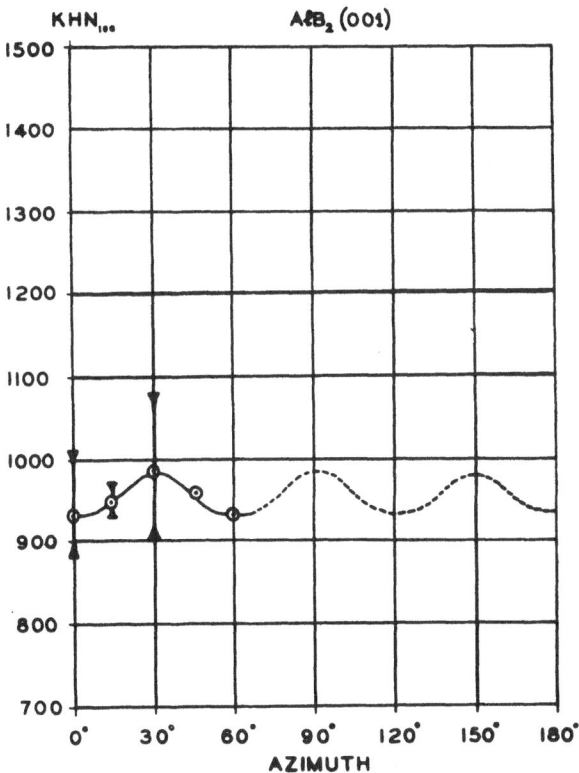

Fig. 3. A plot of the microindentation hardness in the (00.1) plane of AlB_2. The direction of the zero-degree azimuth, defined as parallel to the long diagonal of the indentation, is perpendicular to the direction of the crystallographic a axis. The heavy vertical lines through the first three experimental points show the spread in data for AlB_2.

These data are plotted as curves in Figs. 3, 6, 9, 12, 15, and 18. Although the curves as such are of a somewhat speculative nature, actual experimental points are shown. Since numerical Knoop hardness number (KHN) values can thus be readily derived for any desired azimuth covered, a tabulation of KHN_{100} has been limited to the maximum, minimum, and over-all average for each crystallographic plane investigated. These data are given in Table I.

Table I. Knoop Hardness Numbers (100 g load) for Al, AlB_2, AlB_{10}, α-AlB_{12}, β-AlB_{12}, γ-AlB_{12}, and B

Phase	Plane	Average KHN_{100}	Maximum KHN_{100}	Minimum KHN_{100}	Average variance, %
Al	polycrystalline	19	19	19	± 0.1
AlB_2	(00.1)	960	1085	900	3.1
AlB_{10}	(010)	2610	2785	2490	0.7
	(101)	2785	2850	2725	0.8
	(111)	2550	2760	2350	0.9
α-AlB_{12}	(110)	2445	2500	2390	0.6
	(101)	2380	2450	2320	0.7
	(221)	2210	2250	2170	1.0
β-AlB_{12}	(100)-(010)*	2450	2575	2360	0.8
	(101)-(011)*	2525	2610	2430	0.7
	(201)-(021)*	2870	2990	2660	1.1
γ-AlB_{12}	(100)	2355	2410	2270	0.6
B**	(10.1)	2314	2580	2110	0.5
B	polycrystalline	2460	2480	2415	—

*Polysynthetic twinning.
**Beta-rhombohedral phase.

In order to provide a comparative calibration standard for high-KHN materials, hardness measurements were made on B_4C and SiC and their values compared with those reported in the literature. These data are presented in Table II. Values reported by Cotter [4] for α-AlB_{12} and β-AlB_{12} also are included.

Table II. A Comparison of Observed and Literature KHN_{100} Values for B_4C, α-SiC, α-AlB_{12}, and β-AlB_{12}

Material	Orientation	Average KHN_{100}	Worker
B_4C (molded)	random	2760	Thibault and
α-SiC	(00.1)	2550	Nyquist [5]
B_4C (molded)	random	2755	Cotter [4]
α-SiC	(00.1)	2520	
α-AlB_{12}	random	2433	
β-AlB_{12}	random	2754	
B_4C (uncompacted)	random	2620	This paper
α-SiC	(00.1)	2545	
α-AlB_{12}	random	2380	
β-AlB_{12}	random	2515	

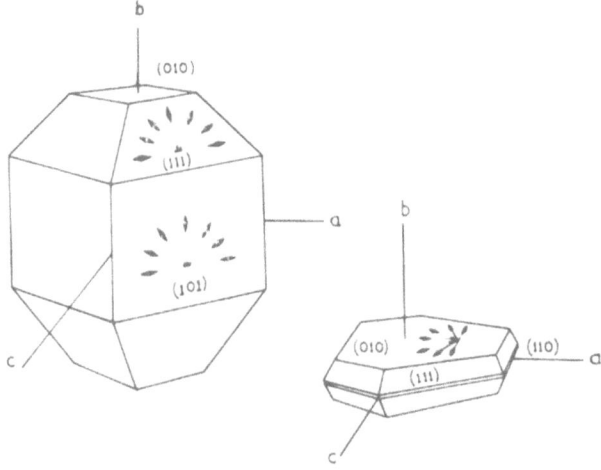

Fig. 4. A crystallographic sketch of commonly observed habits of AlB_{10}. Crystal symmetry is orthorhombic. Microindentation azimuths on the planes investigated are shown.

Fig. 5. A photograph of typical AlB_{10} crystals. Their color is black, with a very bright vitreous luster. Pyramidal and tabular habits are common.

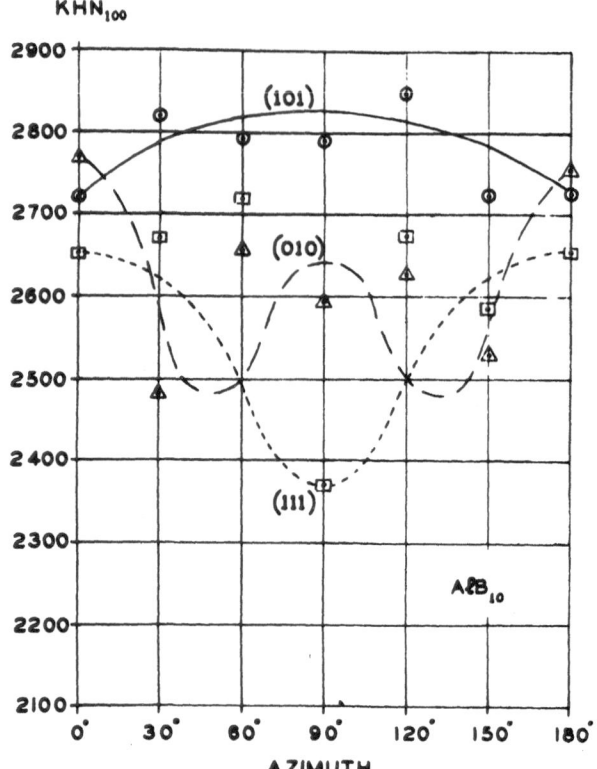

Fig. 6. A plot of microindentation hardness in the (101), (010), and (111)
planes of AlB$_{10}$. The zero-degree azimuth for the (101) plane is parallel to
the trace of the b axis, that for the (010) plane is parallel to the trace of the
a axis. The zero-degree azimuth for (111) lies parallel to the trace of the
plane described by the c and a axes.

Considering the numerous sources of error in this type of work,
the agreement is satisfactory. The difference in value between mold-
ed and uncompacted electric furnace B$_4$C is normal, and the variance
between the aluminum borides as observed by Cotter and this study
can be explained by the greater number of crystallographic planes
and azimuths investigated here.

As mentioned earlier, a detailed compilation of hardness data is
given graphically in Figs. 3, 6, 9, 12, 15, and 18, for AlB$_2$, AlB$_{10}$,
α-AlB$_{12}$, β-AlB$_{12}$, γ-AlB$_{12}$, and B, respectively. In order to make the
presentation more complete, crystallographic sketches are provided
in Figs. 1, 4, 7, 10, 13, and 16. The latter illustrate the most commonly observed crystal habits, the crystallographic planes investigated in this study, and azimuth orientations. In all cases, azimuth
orientation is defined as being in the direction of the long diagonal

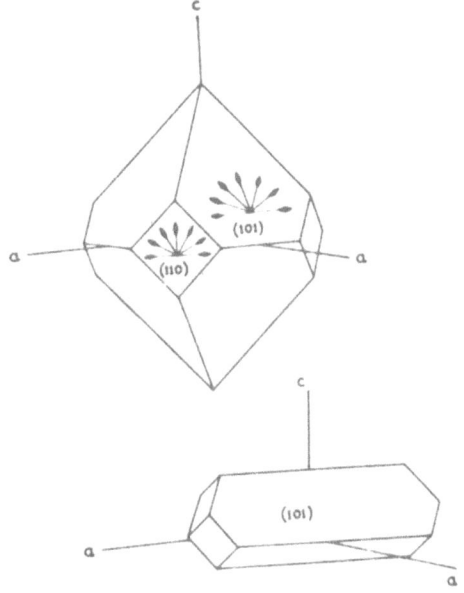

Fig. 7. A crystallographic sketch of typical habits of α-AlB$_{12}$. Microindentation azimuths are shown on the planes investigated. Crystal symmetry is tetragonal (pseudocubic).

Fig. 8. A photograph of typical crystals of α-AlB$_{12}$. Color is black with a vitreous luster.

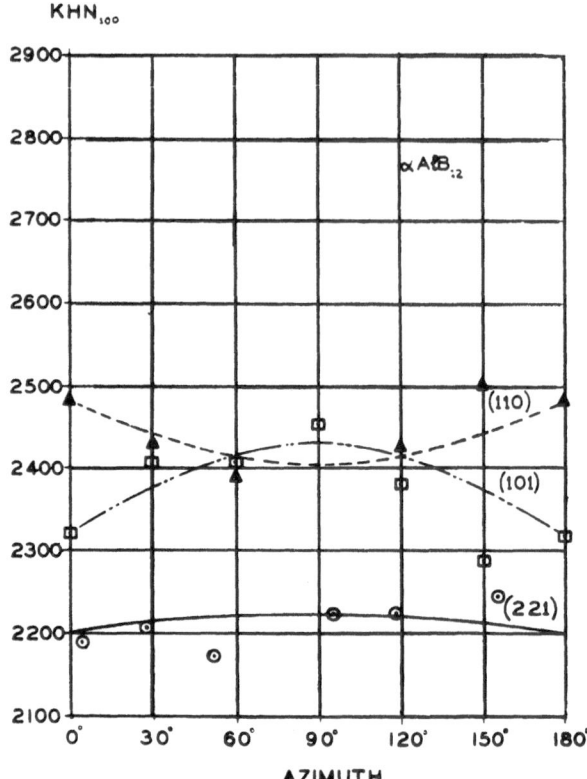

Fig. 9. A plot of microindentation hardness for the (110), (101), and (221) planes of α-AlB₁₂. The zero degree azimuth for (110) is parallel to the direction of the *a* axis. The zero azimuth for (101) lies perpendicular to the plane of the *c* and *a* axes. That for the (221) plane is perpendicular to the direction of the *c* axis.

of the indentation. Vectorial values for indentation hardness thus can be readily derived. Figures 2, 5, 8, 11, 14, and 17 show photographs of typical* crystals of the various compounds in the respective order listed above. Where possible, an attempt has been made to show the development from relatively "ideal" crystal habits to commonly observed, distorted forms.

Sources of Error

The many factors influencing microhardness testing have been analyzed and discussed in detail by Thibault and Nyquist [5] and Bergsman [6]. Essentially, sources of error are both numerous and difficult to control. In this study, an abundance of selected indentation

---

*The boron single crystal is but a fragment with the fortunate development of a large (10,1) face.

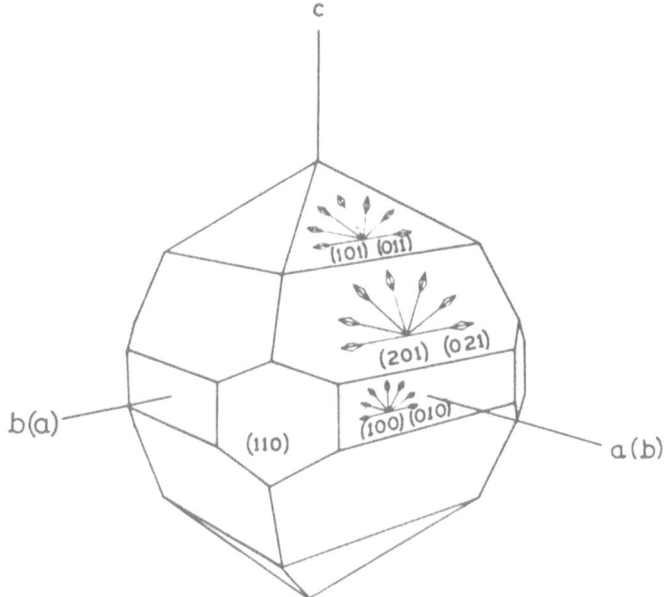

Fig. 10. A crystallographic sketch of a common habit of β-AlB$_{12}$. Microindentation azimuths are outlined on the planes investigated. Crystal symmetry is orthorhombic (pseudotetragonal). The pseudotetragonal nature is emphasized by polysynthetic twinning on the (110) and (1$\bar{1}$0) planes.

Fig. 11. A photograph of typical β-AlB$_{12}$ crystals. Color ranges from yellow to deep amber. Surface luster is vitreous.

data was obtained from a relatively large number of crystals prepared by different methods. Most specimens were studied by two
independent workers. A listing of accountable sources of error is
given in Table III. The most serious source of trouble was found to
be from building vibrations, which tend to cause impact loading. The
latter is the principal cause of indentation fractures. The second
most serious discrepancy was that between different observers for
the same indentation measurement.

Table III. Some Accountable Sources of Error Encountered in
Microindentation Studies and Observed Limiting Values

Sources	Limits, %
Calibration of the optical system	± 1.4
Loading time of indentation.	5.0
Interazimuth orientation.	1.0
Crystallographic azimuth orientation	3.0
Variance between different observers.	2.8
Variance of one observer	1.3

As stated earlier, the hardness curves presented in Figs. 3, 6,
9, 12, 15, and 18 are somewhat speculative. This is easily discern-
ible from the spread of the experimental points used to draw them.
The general trend, however, is believed to be valid. This is illustrated by the limits of variance of the worst set of data, namely, that
for AlB_2. The limits are shown in Fig. 3 by the dark vertical lines
through the first three experimental points. As can be seen, the
general slope of the curve persists. Acceptable microindentations
were difficult to obtain on AlB_2 due to the thinness and flexibility
of the crystals and the excellent basal cleavage.

Additional evidence for the general validity of the hardness data
presented herein can be found in the agreement observed between the
(101) plane of α-AlB_{12} and the (100) plane of γ-AlB_{12}. Due to crystal
structure relationships, the two orientations of the respective mate-
rials are expected to possess similar symmetries and reticular
densities.

Discussion of Results

Four points of interest have been derived from this study: First,
the relatively high degree of hardness of the aluminum borides in
general; second, the consistent correlation between crystal structure
and hardness symmetries; third, the correlation between the degree
of hardness and melting temperatures in the aluminum--boron sys-

tem; fourth, a general contribution to practical knowledge concerning the compounds studied for possible future application to fabrication problems, etc. Since, at present, crystal structures are known for certainty only for aluminum and AlB_2, it is hoped that the data reported here, although incomplete, will be of assistance in resolving some of the problems of structure determination.

With respect to AlB_2, the 60-degree periodicity of hardness anisotropy is consistent with the known crystal structure and reticular density of the (00.1) plane. Hardness on prismatic planes would be expected to be much greater.

AlB_{10} is of interest in that it possesses the greatest average hardness of all phases studied, although only slightly above that of

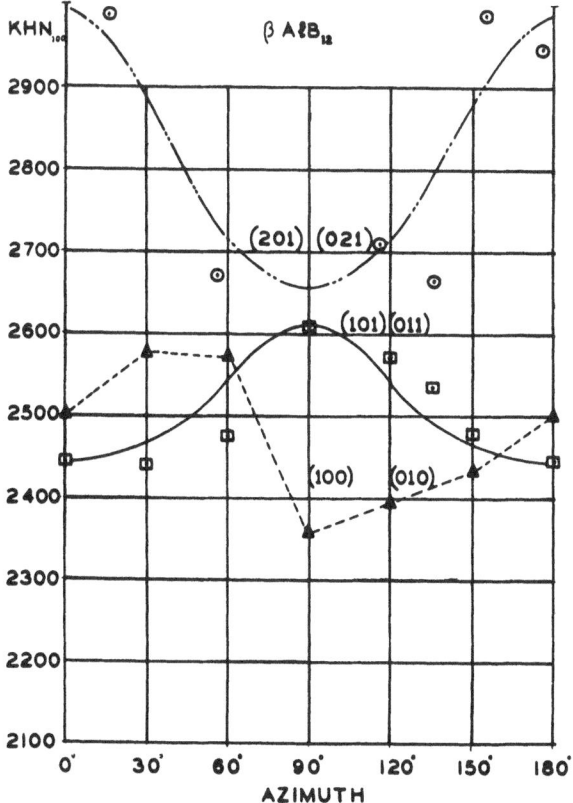

Fig. 12. A plot of microindentation hardness for the (201)-(021), (101)-(011), and (100)-(010) planes of β-AlB_{12} along the azimuths indicated. The zero-degree azimuths for the three planes in the order listed above are: perpendicular to the trace of the (100)-(010) edge, perpendicular to the trace of the c axis direction, and perpendicular to the direction of the c axis.

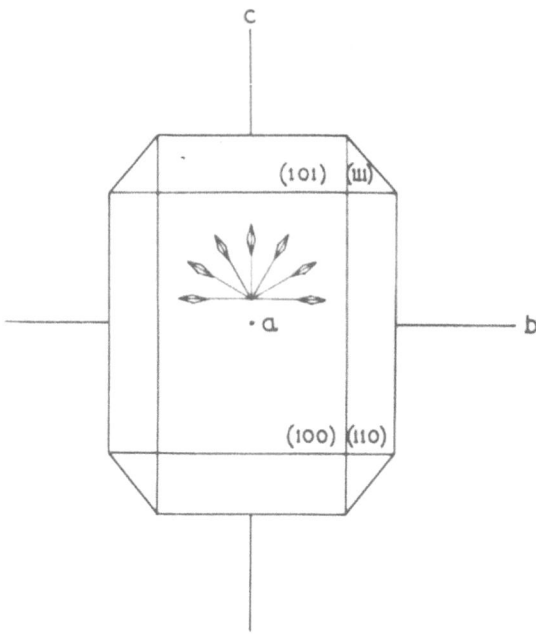

Fig. 13. A crystallographic sketch of a common γ-AlB$_{12}$ habit. Microindentation azimuths on the (100) plane are shown. Crystal symmetry is orthorhombic.

Fig. 14. A photograph of typical γ-AlB$_{12}$ crystals. The latter are frequently intergrown with the α-AlB$_{12}$ phase. Color is black with a vitreous luster.

KHN$_{100}$

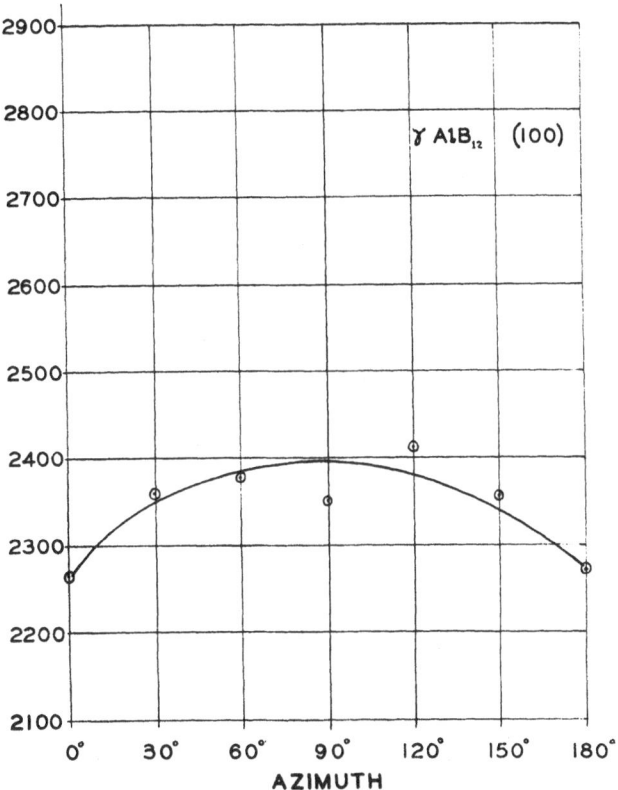

Fig. 15. A plot of microindentation hardness for the (100) plane of γ-AlB$_{12}$. The zero-degree azimuth is perpendicular to the c axis direction.

β-AlB$_{12}$. The latter, in fact, is appreciably harder on the (201)-(021) plane. It is difficult to assess the relative significance of these facts, however, because of the presence of polysynthetic twinning in β-AlB$_{12}$. The hardness anisotropy in AlB$_{10}$ is quite significant both between and within planes. In general, it appears to be compatible with structural symmetry; however, the variation as expressed by the curve for (010) may not be real and seems in need of a modified interpretation.

α-AlB$_{12}$ shows surprisingly little anisotropy, if any, within planes, and no marked variation between planes. What anisotropy does appear to exist, however, is consistent with symmetry requirements.

The highest hardness values obtained thus far are from the (201)-(021) plane of β-AlB$_{12}$. The anisotropy observed in β-AlB$_{12}$ is significant. As previously stated, however, the presence of polysynthetic twinning precludes a simple interpretation of hardness variations.

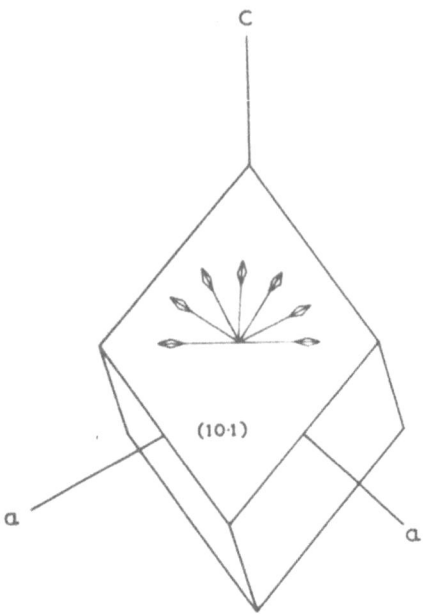

Fig. 16. A crystallographic sketch of an ideal boron single crystal habit. Micro-indentation azimuths of the (10.1) plane are shown. Crystal symmetry is hexagonal (rhombohedral).

Fig. 17. A photograph of the single-crystal fragment of beta-rhombohedral boron used in this study. The flat surface in the plane of the photograph is a natural (10.1) face. Traces of the microindentations made can be seen. Color is black with a bright vitreous luster.

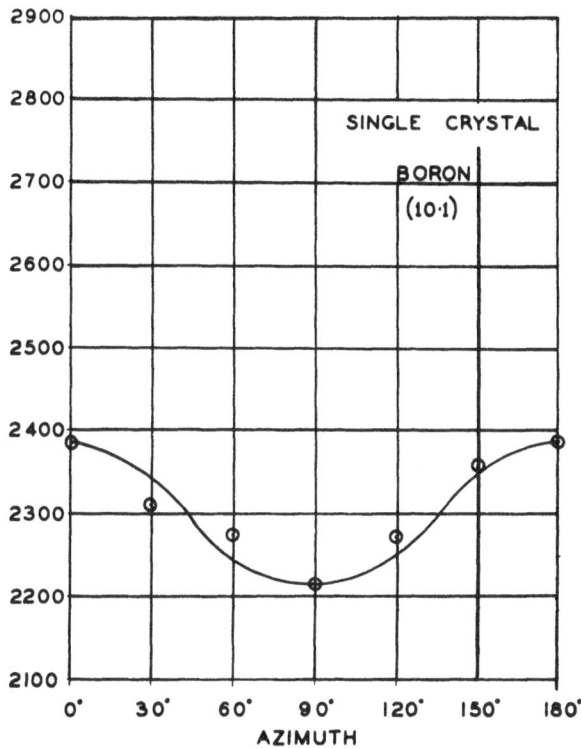

Fig. 18. A plot of the microindentation hardness for the (10.1) face of boron.
The zero-degree azimuth is perpendicular to the trace of the c axis.

The (101)-(011) plane appears to possess a "normal" pattern of anisotropy; however, it is surprising that (201)-(021) shows both the greatest variation and degree of hardness. In addition, the configuration of the (100)-(010) data is quite unexpected, but seems to be real since it has been observed on two different crystals by two observers. No explanation is offered. It perhaps should be mentioned here that AlB_{10} crystals often are found both intergrown with and included within β-AlB_{12} crystals. The specimens used in this study, however, were pure phases.

Only the (100) plane of γ-AlB_{12} was investigated in this study. γ-AlB_{12} crystals generally are found as intimate intergrowths with α-AlB_{12} and can be distinguished visibly only by the frequent appearance of the rectangular (100) face when the γ-AlB_{12} phase predominates. The two phases are closely related structurally, the (100) plane of γ-AlB_{12} being common to (101) of the alpha phase. This similarity is clearly illustrated by the almost identical hardness anisotropy consistent with the symmetrical relationship between the two.

Due to the availability of only one specimen of single-crystal boron* and its physical configuration, only the (10.1) plane could be investigated. Extensive measurements were carried out, however, and the data are probably the most thorough of all reported here. As can be seen from Fig. 18, the variation of hardness corresponds to symmetry requirements for this orientation.

The phase vs average hardness and phase vs melting temperature diagrams of Figs. 19 and 20, respectively, are presented to show qualitatively the correspondence between these two properties of the several phases in the aluminum--boron system. Melting temperature determinations were made by placing single crystal specimens** in a hot-pressed boron nitride crucible. The crucible was heated by an

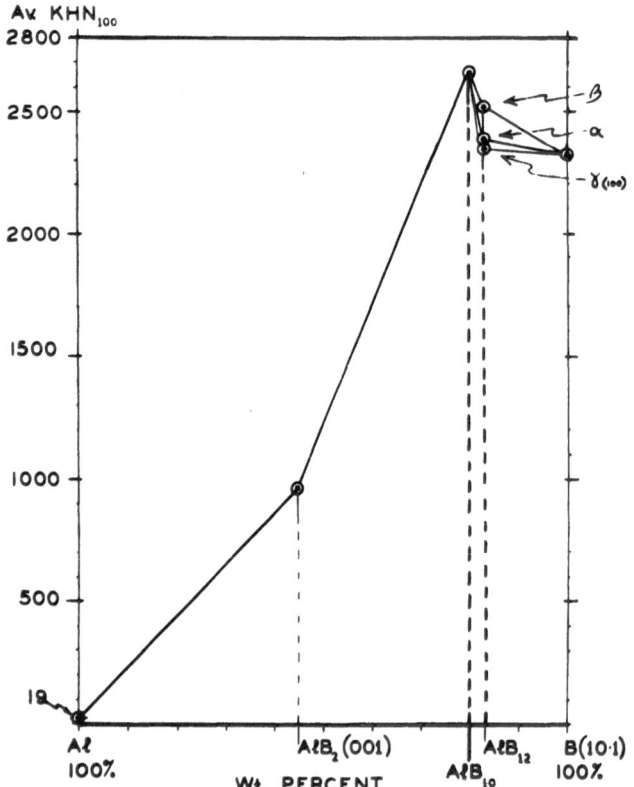

Fig. 19. A phase—hardness diagram for the aluminum—boron system. The hardness is expressed in terms of the average Knoop hardness number for a 100-g indentation load.

---

*High-temperature, beta-rhombohedral variety.
**Polycrystalline aluminum and boron were used.

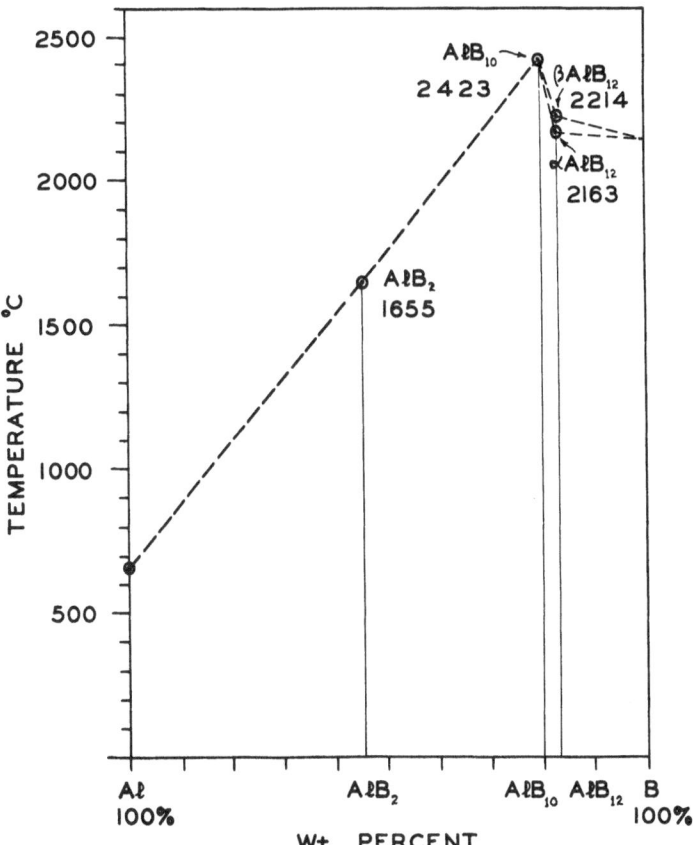

Fig. 20. A phase—melting point diagram for the aluminum—boron system, with the exception of γ-AlB$_{12}$. (Temperatures shown are averaged values. Limits of reproducibility are ± 50°C.) Note the general similarity with Fig. 19.

electrical resistance heating ribbon. Temperatures were read with an optical pyrometer and are believed to be accurate to ± 50°C*. In order to retard the loss of aluminum from the boride structures at high temperatures, measurements were made in a purified argon atmosphere maintained at the following series of pressures: 29, 40, and 51 psi. Melting-temperature curves were extrapolated to the point of essentially no change with increasing pressure, and values taken at this point. Reactivity with the crucible material was a problem. The latter was minimized, however, by rapidly reaching the melting temperature.

---

*Gold and platinum were used as calibration standards.

Acknowledgment

The authors would like to express their thanks to J. E. Tydings for preparing all of the drawings used in this article.

References

1. Kohn, J. A., Gaulé, G. K., and Giardini, A. A., Proc. of Second Army Sci. Confr., (classified) West Point, N.Y. (June 1959).
2. Kohn, J. A., Katz, G., and Giardini, A. A., Zeit. f. Krist. 111 (1958) 53-62.
3. Kohn, J. A., these Proceedings, p. 75.
4. Cotter, P., Am. Min. 43, No. 7-8 (1958) 781-784.
5. Thibault, N. W., and Nyquist, H. L., Trans. Am. Soc. Metals 38 (1947) 271-330.
6. Bergsman, E. B., ASTM Bull., No. 176 (Sept., 1951).

OPTICAL AND ELECTRICAL PROPERTIES OF BORON AND POTENTIAL APPLICATION

G. K. Gaulé, J. T. Breslin, J. R. Pastore, and R. A. Shuttleworth*

The material investigated was mostly of the beta-rhombohe-dral form, and of the highest purity available at present. When necessary, single crystals or crystals having a slight mosaic structure were used. The surface structure of these crystals is shown, and an explanation given of the growth mechanism. Among the samples investigated there was a wide spread of impurity concentrations as determined by spectrochemical and vacuum fusion analysis. The spread in electrical properties was comparatively small, but there were few significant exceptions. The resistivity values were in the megohm-centimeter range; the Seebeck coefficient was approximately 500 $\mu v/deg$; the classical semiconductor contact phenomena, such as rectification and the photovoltaic effect, were not found. The variation among infrared absorp-tion spectra is much larger than that of the electrical values, and the material could be grouped into four distinct classes, A, A', B, B', according to the spectra. Transitions between classes could be introduced by removal or introduction of impurities. Class A' represents the most transparent ma-terial. An optical band gap of 1.56 ev and an absorption coef-ficient of only $10 \, cm^{-1}$ were found in the best material of this class, which was also the purest, with impurities at or below the 10 ppm range. Such material could be used for heat- and corrosion-resistant windows in the 1-8 μ region of the infrared. The high temperature coefficient of the ma-terial (ionization energies are approximately 0.6 ev at room temperature, and increase up to 1.5 ev at high temperatures) leads to applications in thermistors, etc. The main obstacle to other semiconductor applications seems to be the large concentration of traps in present-day boron, which becomes evident in the long relaxation times observed in photocon-ductivity and certain magnetic experiments.

---

*U.S. Army Signal Research and Development Laboratory, Fort Monmouth, N.J.

I. *Introduction*

The essential semiconductor properties of "pure" crystalline boron, namely the high resistivity, the large negative temperature derivative of the resistivity, and the large Seebeck coefficient were reported by Weintraub as early as 1909 [1]. In the same year, use of boron in thermocouples and similar technical devices was suggested by the same author [2]. Photoconductivity was found in 1934 [3], and infrared absorption curves, also confirming the semiconductor nature of the material, were published in 1953 [4]. In spite of this agreement on the general nature of boron material, large discrepancies can be found in the older literature among the characteristic quantities reported, especially among values for the band gap. Part of these discrepancies may be explained in terms of the unusually large concentrations of impurities which are found even in carefully purified, high-resistivity material, as shown in section III of this article. The main reason for the discrepancies, however, seems to be the three (at least) modifications in which boron can crystallize, depending on the conditions of formation, in particular the temperature. Much of the material investigated was apparently a mixture of two or more modifications, or else, not dense and crystalline at all [5]. Structure data and temperature ranges of formation for the established modifications have been determined and compiled by Hoard [6] and Hoard and Newkirk [7]. Work reported in this article has focused on the investigation of crystalline material obtained from a melt. Such material is always dense and consistently of the high-temperature or beta-rhombohedral modification.

II. *Crystallization*

To grow single crystals of beta-rhombohedral boron, a pool of melted boron was created on top of a large piece of boron, using induction heating of part of the material by a highly concentrated radio-frequency field. This crucible-free method [8] has been successfully applied to pull single crystals of germanium and silicon. As yet the method has failed in the case of boron, probably due to the tendency of boron to crystallize in platelets and to crack upon cooling. It was found that a small platelet can be grown using a modification of the process; the center part of the melt is slightly supercooled and momentarily touched with a heat sink. The growth mechanism appears to be related to the dendritic mechanisms described, for example, by Saratovkin [9], an assumption which is supported by the macroscopic (Fig. 1) and the microscopic (Fig. 2) growth patterns found on such platelets. Similar patterns were often observed on the top face of material produced and melted in vacuo by U.S. Borax & Chemical Corp. [10]. Figure 3 shows such material with a large rhomboid pattern. The groove around the rhomboid was

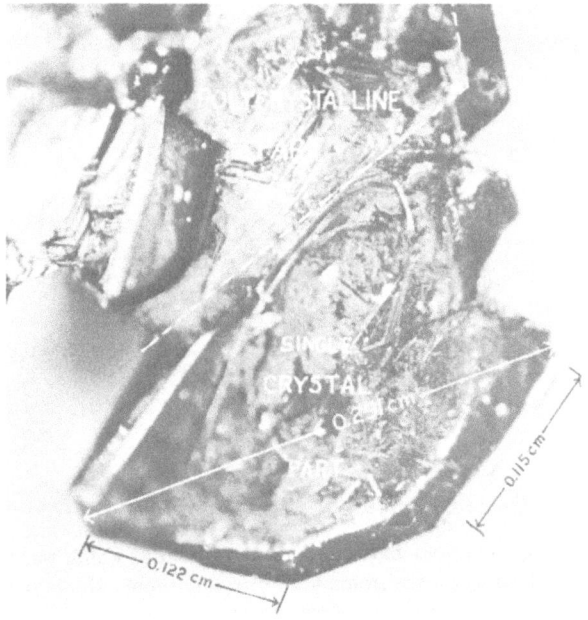

Fig. 1. Single crystal grown from supercooled melt (81030).

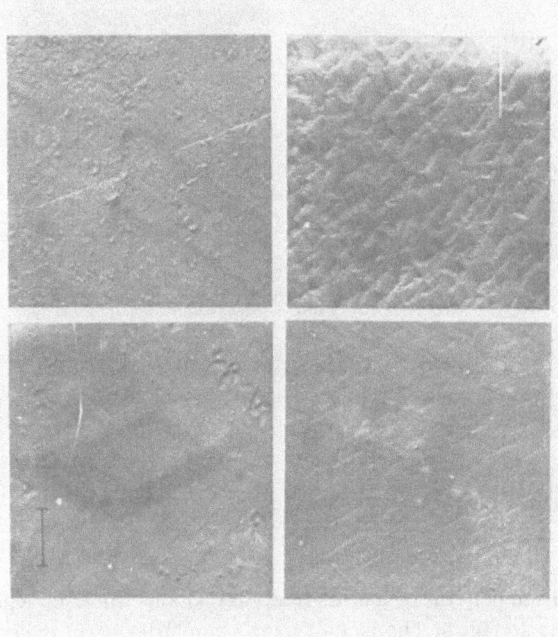

Fig. 2. Microscopic growth patterns on single crystal of Fig. 1, as revealed by electron microscope. Scale indicates 1μ.

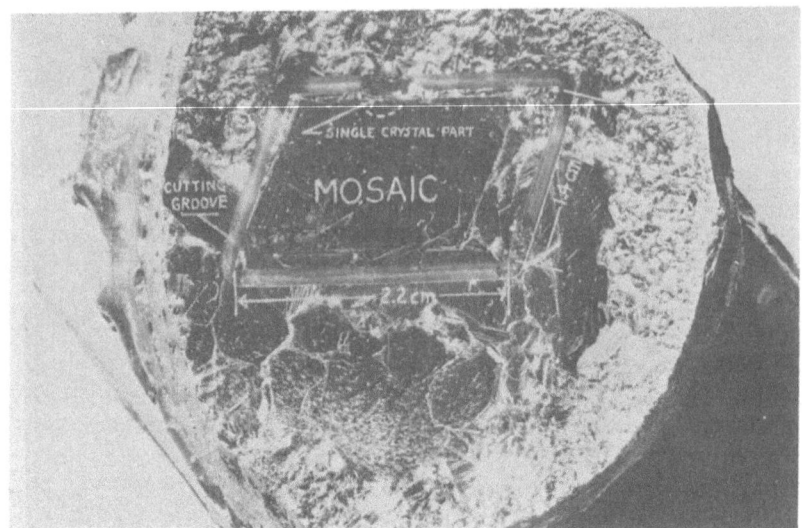

Fig. 3. Large rhomboid pattern on top of material which was melted in vacuo. Cutting groove stems from ultrasonic tool, which later separated the center piece completely from the host material (81213).

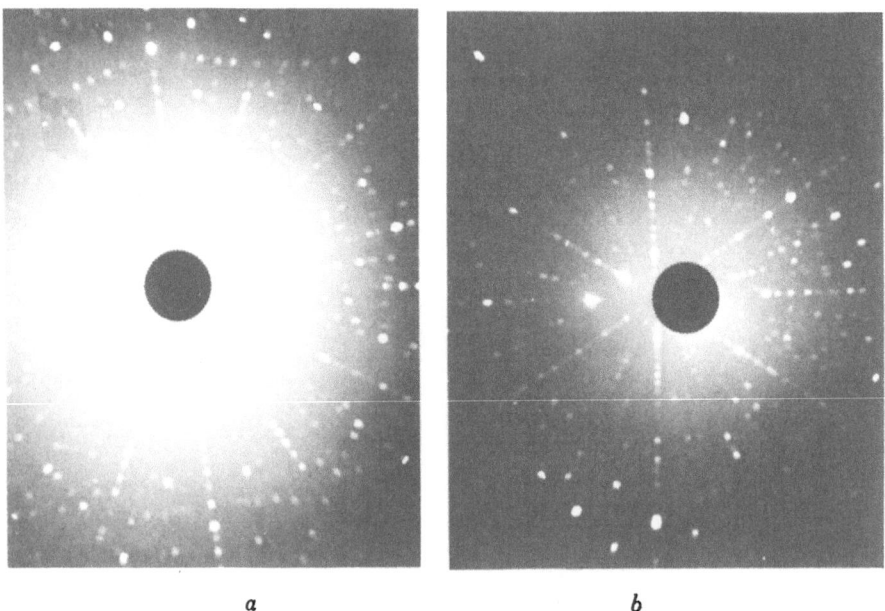

a   b

Fig. 4. Laue patterns of the single crystal(a) and "mosaic" regions (b) indicated in Fig. 3. The lack of some symmetries in (b) indicates the existence of mosaic blocks.

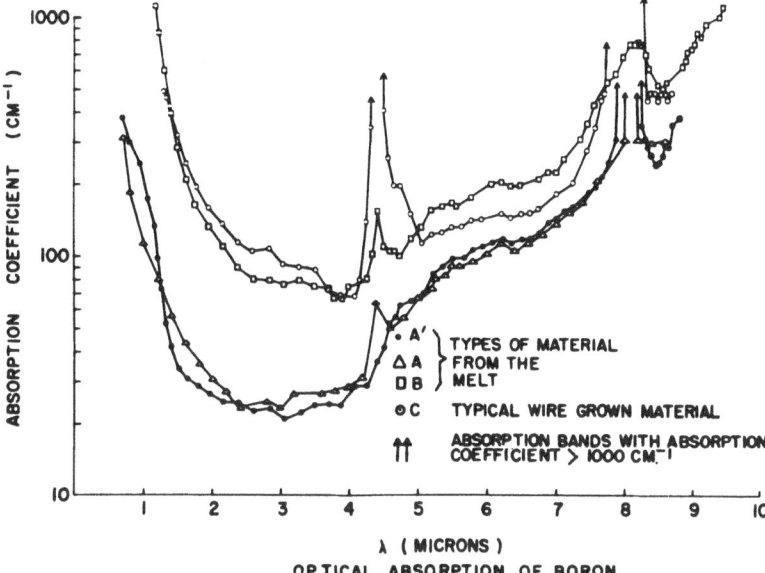

λ (MICRONS)

OPTICAL ABSORPTION OF BORON

Fig. 5. Typical absorption spectra. Classes A on the one hand, and B and C on the other hand differ with regard to the absorption edge. Material without the 4.4 μ absorption band is, in addition, designated with a prime (A'). The A-curve was taken from [11], the A'-curve from sample 90211, which was obtained from the C material 90104 through float-zoning in vacuo. The B-curve was taken from the single crystal 81030.

cut with an ultrasonic tool, which was later used to free the suspected single-crystal part completely from the polycrystalline host material. Laue back-reflection patterns were then obtained at several points, and it was found that only a small region was monocrystalline (Fig. 4a), whereas the bulk has a mosaic structure as indicated by the lack of symmetry in the Laue pattern (Fig. 4b). It might also be inferred from this pattern, however, that the tilt among the mosaic blocks is very small and of negligible influence on most of the electrical and optical properties. Mosaic material was used for all the experiments where large samples were necessary.

III. *Optical Properties and Chemical Composition*

The investigation of the infrared absorption edge of a number of samples of beta-rhombohedral crystalline boron led to a grouping into two classes. Class A comprises material with an absorption edge at approximately $0.8\,\mu$ and Class B material with an absorption edge at approximately $1.2\,\mu$. (Fig. 5). According to Moss [12], the steepest part of each curve was taken for the preliminary determination of these values. No significant difference between the two

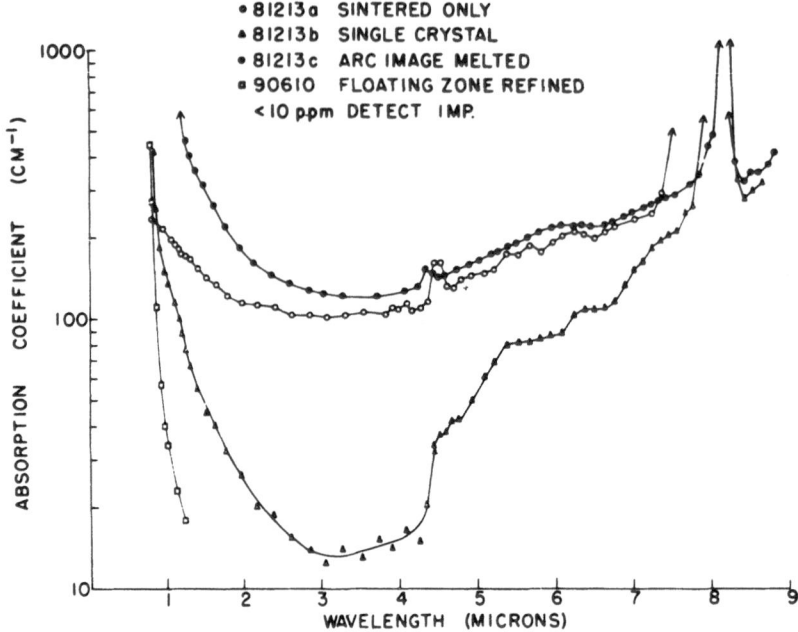

Fig. 6. "Window" properties of boron, depending on preparation and purity. Reduction of the impurity level below 10 ppm makes absorption edge steeper. Host material and single crystal 81213 can be seen in Fig. 3.

classes with regard to impurity concentration, crystal structure, or electrical properties could be found. Experience shows, however, that crystallizing in a vacuum leads consistently to class A material, whereas crystallizing in an impure inert atmosphere leads occasionally to class B material. Class C denotes wire-grown material, the structure of which is often uncertain, as mentioned in the introduction. The absorption spectrum of this material is similar to that of class B. Figure 6 illustrates the influence of purity and treatment on the absorption. Sample 90610 is highly purified, whereas the other material is of only moderate purity, as shown in detail in Table I. Sample 81213a represents the parent material for 81213b and c. Since the parent material is only sintered, it is rather opaque. Melting in a vacuum led to the single crystal shown in Fig. 3, with a rather steep edge. On the other hand, melting and crystallizing the same material in an arc image furnace led to rather opaque class B material, probably due to impurities in the inert atmosphere used (curve 81213c, Fig. 6).

The investigation of the photoconductivity of class A material revealed a spectral response with a sharp peak at approximately $0.8\,\mu$, very close to the absorption edge as defined above (Fig. 7).

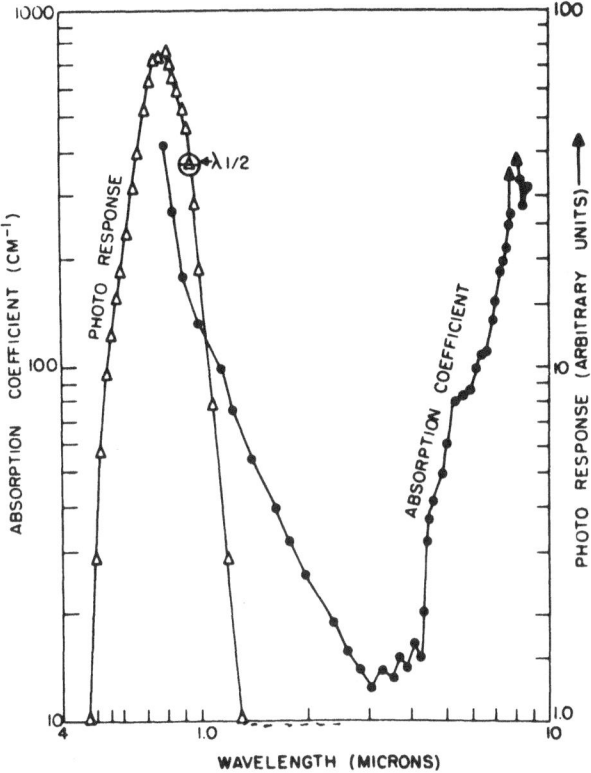

Fig. 7. Photoconductivity in class A* material (single crystal 81213b). The photo response reaches maximum at a relatively small absorption value and decreases rapidly as absorption increases. Some photoconductivity is noticeable up to 2.5 μ.

A similar peak was found for crystalline material by Lagrenaudie [4]. Moss's $\lambda\frac{1}{2}$ value [12]--a longer wavelength at which the photo response has decreased to one half its maximum--lies at 0.98 μ in Fig. 7. Theory, and also absorption data obtained from boron films [12,13], indicate that the absorption coefficient goes to values 10^5 cm^{-1}, and higher, as the wavelength goes below the absorption edge. Light of that range, which penetrates less than 0.1 μ into the boron, yields only a small photo response, as shown in Fig. 7. This is in contrast to the behavior of silicon [14]. These and other phenomena which are discussed in section IV, suggest a very short diffusion length for carriers in boron, so that almost all electrons and holes generated near the surface recombine before they can make a contribution to the photocurrent. A small, but measurable, photo response was found for wavelengths longer than that of the peak, up to 2.5 μ (see dashed part of curve in Fig. 7). The decay of the photocurrent

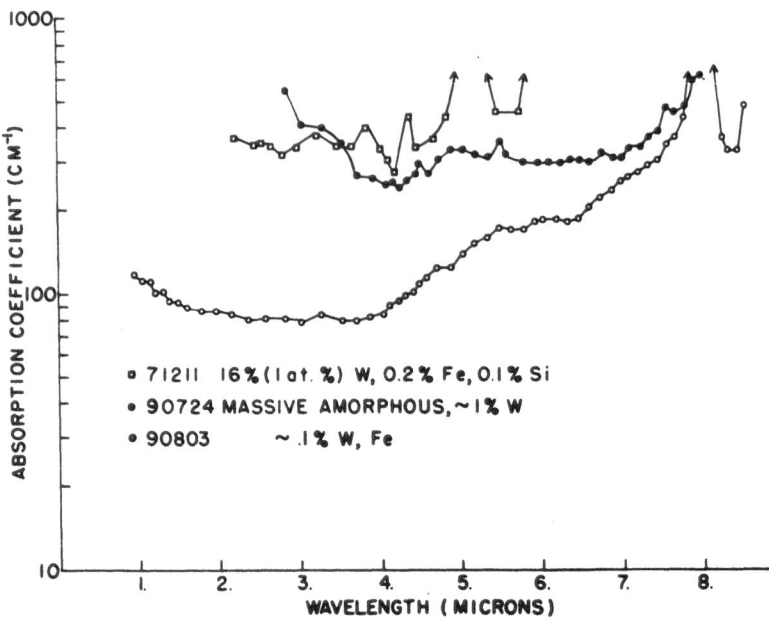

Fig. 8. "Opaque" material, containing various amounts of W. Sample 90724 is not crystalline and seems to have a band edge at a wavelength longer than that for crystalline material.

after the light is turned off does not follow a simple exponential law; the decay rate decreases with time. The initial slope of the decay curve, however, can be characterized by a "time constant." A typical value after weak illumination is 20 seconds. After strong illumination (100-w tungsten lamp at 20 cm) the value is only a fraction of a second. These observations indicate a high density of traps [15] in the material. In the case of strong illumination most of the traps are saturated and the direct, fast recombination processes are predominant.

Typical absorption curves, as plotted in Fig. 5 and 6, consistently show an absorption band at about 8μ, probably a lattice vibration band. Part of these curves show absorption bands of varying strengths at about 4.4μ. This band is apparently caused by the "B–H stretching vibration" of hydrogen atoms bound to boron atoms in the lattice. The analogous bands for boron–hydrogen compounds are at approximately 4μ [16,17]. The 4.4μ band and most of the hydrogen, as shown in Table I, are removed when the material is melted in vacuo. Typical examples are shown in Fig. 5, where 90104 represents the parent material for sample 90211, which was float-zoned in a vacuum, and in Fig. 6, where 81213a and b represent analogous cases.

It is convenient to classify material not having the $4.4\,\mu$ absorption by adding a prime (') to the symbols A or B, these latter designating the position of the band edge. Samples of the four possible cases, namely A, A', B, B', have been found.

The role of tungsten as an impurity is also qualitatively understood. X-ray diffraction studies have shown that the tungsten atoms are located at nonrandom sites having relatively high symmetry. In this arrangement, the tungsten atoms seem to supply a large number of electrons to the conduction band, rather independent of temperature. As Fig. 8 indicates, the material becomes more and more opaque with increasing tungsten concentration. Also, the electrical properties change toward that of a metal, as discussed in section IV. In contrast to tungsten, the effect of comparable concentrations of iron, titanium, silicon, etc., on the optical and electrical properties is very small, as shown in Table I.

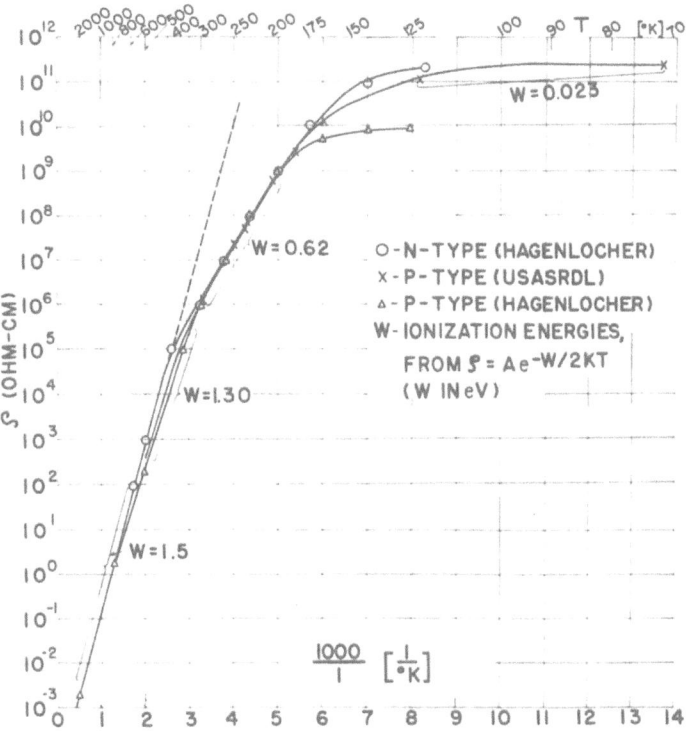

Fig. 9. Typical $\log \rho$ vs. $1/T$ curves for high-resistivity material. "Mosaic" material, class A', from 81213 was used for the USASRDL curve. The other curves are taken from [19] for comparison.

Table I. Chemical Composition and Physical Properties

Sample number	Figure	Structure and preparation	O	H	N	C	Si	Fe	W	Ca	Others	Optical properties (Class)	Type	ρ	Magnetic properties
71211	8	Vacuum-melted with tungsten	--	--	--	--	$10 \cdot 10^2$	$20 \cdot 10^2$	$16 \cdot 10^2$	10	Ni 100	low transmission	N	~1	--
81030	1,2,5	Beta-rhombohedral. Single crystal. Pool melt, mostly from 81213	--	--	--	--	$10 \cdot 10^2$	100	--	--	Ta $3 \cdot 10^3$ Mn 500	B	P	10^6	--
81213b	3,4,6 7,9	Beta-rhombohedral. Mosaic, vacuum-melted	56	13	<5	830	500	500	nd	--	Mn 300 Mg 200	A'	P	10^6	CRM negative
90104	5	Polycrystal. Deposited on Ti filament	--	--	--	--	5	nd	--	10	Ti $1 \cdot 10^4$ Cu 1	C	P	10^6	--
90206	5	Polycrystal. Deposited on Ta filament	14	403	<25	100	200	nd	nd	nd	nd	C	--	--	--
90211	5	Originally 90104 then float-zoned in vacuum	--	--	--	--	5	nd	--	10	Ti $1 \cdot 10^4$ Cu 1	A'	P	10^6	--
90518		Polycrystal. Deposited on boron rod and float-zoned four times	--	--	--	--	200	50	--	nd	nd	B'	P	10^6	--
90518a		90518 before treatment in O_2 one hour at 700°C	$3.5 \cdot 10^3$	44	<60	290	--	--	--	--	--	B'	P	10^6	--
90625	5	Polycrystal. Deposited on Ti filament	47	415	<7	$10 \cdot 10^2$	50	nd	nd	1	Ti $1 \cdot 10^4$	C	--	--	--
90724	8	Massive amorphous. Deposited on W filament	--	--	--	--	5	nd	$10 \cdot 10^3$	1	Mg 3	low transmission	--	--	--
90803	8	Polycrystal. Deposited on W filament then arc-melted	--	--	--	--	--	approx $10 \cdot 10^2$	approx $10 \cdot 10^2$	--	--	low transmission	--	--	--
91110	5	Polycrystal. Vacuum-melted	700	50	200	--	500	$18 \cdot 10^2$	--	300	Mn 500 Al 150	A'	P	10^6	--
81213a	6	Sintered in vacuum. Bottom of bulk from which 81213b was obtained	--	--	--	--	$10 \cdot 10^2$	$30 \cdot 10^3$	nd	500	Mg 1000 Mn 2000 Ni 500	low transmission		--	CRM negative
81213c	5,6	Same material as 81213a. Melted in impure argon	--	--	--	--	--	--	--	--	--	B	--	--	--
906 0	5,6	Polycrystal. Deposited on Ta filament then float-zoned 3 times	--	--	--	--	10	10	nd	10	Al 10 Cu 10	A'	P	10^6	CRM negative

ρ is the resistivity in ohm-cm (order of magnitude); CRM is the relative change in resistivity after magnetic treatment; O, H, N are determined by vacuum fussion, C by conductometric, and other elements by spectrochemical analysis; nd—not determined; (- -) not tested.

Up to 50 ppm oxygen was found in material of classes A' and C. The concentration was brought up to 3500 ppm in a B' sample by heat treatment at 700°C in oxygen (90518a in Table I). The oxygen content could not be related to optical data; it should be observed, however, that very little information is available on the absorption beyond 10μ. Appreciable transmission from 12.5 to 15μ was observed in class A' samples in this study. Less transmission was found in class A material by Spitzer and Kaiser [11].

Preliminary results indicate that material of all classes can contain appreciable amounts of carbon.

Boron single crystals are apparently birefringent. Polarization phenomena were observed when a crystal was rotated about the [10$\bar{1}$1] axis under illumination with polarized light of approximately 1μ wavelength.

IV. *Electrical Properties and Potential Applications*

To gain insight into the potentially useful semiconductor properties of beta-rhombohedral boron, a qualitative investigation of a number of metal--boron pressure contacts was made. Class A and class A' materials, which are consistently of a resistivity of 10^5 to 10^6 ohm-cm, p-type, were used. When metals with a relatively small work function, such as copper or zinc, were used, the Seebeck coefficient was high--in the neighborhood of 500μ v/deg. These results are compatible with calculations presented by Ioffe [18] if one assumes that the Fermi level lies not much below the center of the energy gap. This assumption, in turn, is supported by the form of the temperature dependence of the resistivity, which is presented in Fig. 9 and discussed below. All experiments designed to discover rectifying effects in metal--boron contacts had negative results. Small nonlinearities found at low temperatures (80°K) could be explained in terms of a local heating effect. Area contacts were made by baking a silver or gold paste onto the boron surface. These contacts were also consistently ohmic. The resistance of mechanically sound area contacts was consistently negligible in comparison to the bulk resistance. Furthermore, no photovoltaic effect could be detected in any of the contacts investigated. These findings are somewhat atypical for a high-resistivity semiconductor. As in the case of another uncommon phenomenon, namely the very slow decay of photoconductivity (section III), a high trap density is suggested for the explanation. The traps may prevent the formation of a depletion layer which is thick enough to yield a diode characteristic for the contact. Experiments with low-resistivity material are necessary to clarify this point further.

In this study, investigations of the resistivity and its temperature dependence were limited to high-resistivity, p-type material. All samples from classes A, A', B, B', and some from class C, seem to fall into this category, in spite of the high concentration of impurities in general and the wide spread among the chemical compositions, as shown in Table I. The concentration of acceptors, however, may only be between 0.01 and 0.1 ppm [19]. Figure 9 shows a typical curve compared with data by Hagenlocher [19] on high-resistivity p- and n-type material. It is seen that the activation energies are in the neighborhood of 0.6 ev. The activation energy of 1.5 ev for the intrinsic range is compatible with the value of 1.56 ev calculated from the absorption edge (section III).

It was shown in section III that crystalline boron can accommodate a large concentration of gaseous impurities. This is also reflected in resistivity measurements on class A material, heat-treated at 400°C in different gas atmospheres by Medcalf [20]. Oxygen and nitrogen increased the resistivity by approximately 200%; hydrogen decreased it by 35%.

If the concentrations of metallic impurities typical for class A or B material and listed in Table I are sufficiently exceeded, medium resistivity material is obtained. Results of heavy doping with metals and with carbon are given by Hagenlocher [19], with phosphorous by Greiner and Gutowski [21], and for metallic impurities in tetragonal boron by Shaw et al. [22]. As discussed in section III, the most striking change in the properties of boron is produced by the introduction of tungsten. Depending upon the tungsten concentration, the resistivity drops to 10 ohm-cm or even less, the temperature derivative of resistivity becomes very small and positive (metallic), and the Seebeck coefficient becomes very small. The material is consistently n-type. This kind of material may find some application in resistors.

The large intrinsic and extrinsic activation energies of high resistivity boron, in combination with the convenient contact properties, make this material promising for application in thermistors. The performance of a thermistor [23,24] is characterized by α, the relative change of resistance (and, therefore, resistivity of the thermistor material) per degree change of temperature. In general, α is strongly dependent on temperature, as may be deduced from the approximate formulas given below.

$$\alpha = \frac{1}{\rho}\frac{d\rho}{dT} = \frac{d\ln\rho}{dT} = -\frac{d\ln\sigma}{dT} = -\left(\frac{1}{T}\right)^2\frac{d\ln\sigma}{d(1/T)} \qquad (1)$$

$$\ln\sigma = \ln\sigma_0 - \frac{2\alpha}{1/T}, \quad \text{for } \alpha = \text{const.}, \qquad (2)$$

$$\sigma = \sum_{n=0}^{n=n} \sigma_{\infty n} e^{-W_n/2kT}; \quad \sigma_{\infty n+1} \ll \sigma_{\infty n}; \quad W_{n+1} \ll W_n; \qquad (3)$$

when: $\sigma \approx \sigma_{\infty l}\, e^{-W_l/2kT}$, $\alpha \approx \dfrac{-W_l}{2k}\left(\dfrac{1}{T}\right)^2$ ⠀⠀⠀⠀⠀ (3a)

$$N_\blacksquare \approx \left(\dfrac{\sigma_{\blacksquare\infty}}{e\mu}\right)^2 \dfrac{1}{C_0}\;;\quad C_0 \approx 2.5\cdot 10^{19}\ \mathrm{cm}^{-3}.$$ ⠀⠀⠀⠀ (4)

For one thing, formula (3a) contains the factor $1/T^2$ which becomes very large for low temperatures. On the other hand, first the intrinsic

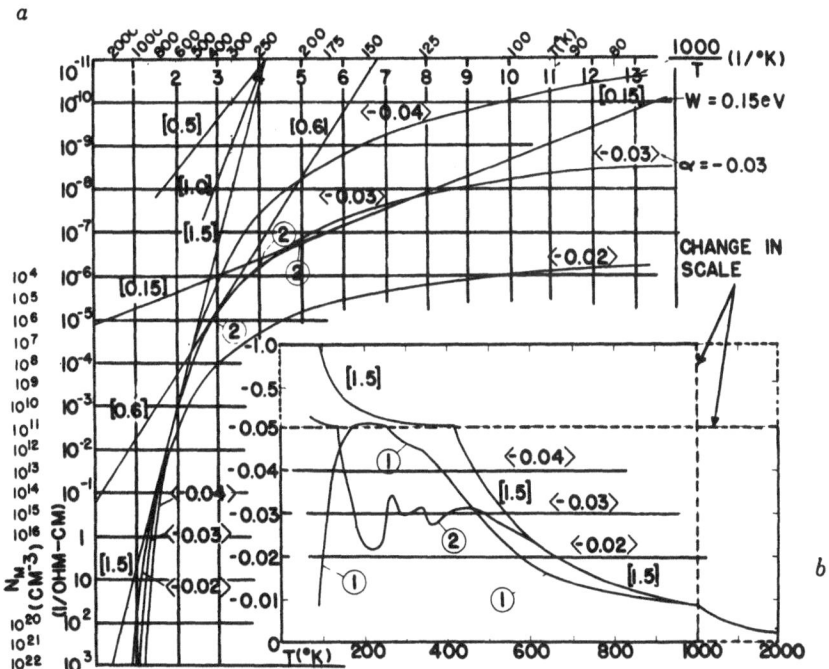

Fig. 10a. $-\log \sigma$ ($\asymp +\log \rho$) is plotted versus $1/T$. Hereby the scales are adjusted to those of Fig. 9 so that comparisons between empirical and "ideal" curves may be made. $<-0.03>$ represents the "ideal" curve of a thermistor with an $\alpha = -0.03$ for all temperatures. Curve $<-0.03>$ is closely approximated by the curve 2, which is obtained by graphical construction under the assumption of class A' material with a few (10^6 cm^{-3}) shallow (activation energy 0.15 ev) acceptor levels added.

Fig. 10b. α is plotted as a function of temperature. It is seen that curve 2 leads to a reasonably flat characteristic. This is also true, but to a lesser degree, for the empirical characteristic 1 which is calculated for the class A' material of Fig. 9. Curve [1.5] represents the case of consistently intrinsic material, with extremely high α values at low temperatures.

activation energy W_0 and then a sequence of extrinsic activation energies W_1, W_2, etc., may become predominant as the temperature is lowered. Since the quantities W_l form a monotonically decreasing sequence and successively become factors in (3a), the $1/T^2$ influence on the temperature dependence of α can be nearly canceled if the activation energies of the impurities, and their concentrations, N_1, are properly chosen. This is illustrated in Fig. 10a. Using formula (2), one determines for example the hypothetical curve <-0.03>, belonging to an α which is strictly constant, namely −0.03 (equivalent to −3% change in resistance) for all temperatures below 900°K. It is significant that the form of the empirical curves taken from class A' material, as shown in Fig. 9, comes rather close to the form of this hypothetical curve. The knee introduced by the onset of extrinsic conduction prevents the α values from becoming very high. This is shown quantitatively in Fig. 10b, where curve 1 describes the temperature dependence of α for class A' material. It is seen that α stays between −0.02 and −0.04 for a wide temperature range. Furthermore, curves 2 in Figs. 10a and b indicate that the "ideal" curve <−0.03> could be approximated more closely, and the α vs. temperature characteristic made still "flatter," if, for example, $10^6 \, cm^{-3}$ additional acceptor levels with activation energies of approximately 0.15 ev would be introduced into class A' material. Therefore, the manufacture of boron thermistors with "flat" characteristics seems possible. Another advantage of the use of class A' boron seems to be that it is crystalline and corrosion resistant and may operate more reliably in a vacuum or in a corrosive environment than the usual thermistor material which is sintered from a powder.

Several thermistors were built using bars of class A' material onto which silver contacts were baked (Fig. 11b). The curves (Fig. 11a) were obtained under self-heating conditions using the thermocouple in the housing to determine the temperature. α values calculated from the self-heating curves are in agreement with curve 1 of Fig. 10b. The performance of a thermistor under self-heating conditions can be estimated from the knee in the curves (Fig. 11a). It is dependent not only upon α but also upon the thermal insulation of the thermistor material against the environment. No attempts were made to optimize these conditions. Therefore technical refinements may considerably improve the performance.

V. *Magnetic Effects*

A rather unusual magnetic effect was found in some of the class A' samples. Subjecting the crystals to magnetic pulses (5000 oersted, 100 seconds, for example) decreased the resistivity. A decrease of 25% after ten such pulses was typical. The original resistivity value reappears only very slowly, unless the material is heated to 200°C and allowed to cool to room temperature. No appreciable variations

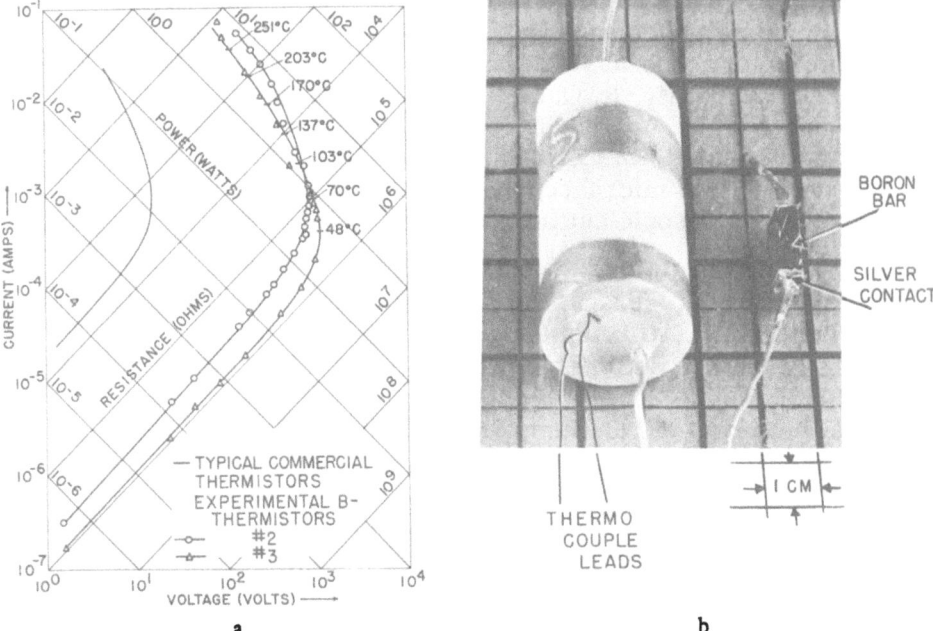

a   b

Fig. 11. Experimental boron thermistors. Mosaic material (81213) was used. The characteristics were obtained under self-heating conditions. Hereby the thermocouple served to determine the boron temperature.

in the effect were observed when samples with different iron concentrations were investigated. Also, the orientation of the crystal in the field seems to play little or no role. These investigations are being continued.

VI. Conclusions

The electrical and optical properties of high-resistivity, beta-rhombohedral boron are remarkably uniform in spite of wide variations in the chemical composition. Therefore, many effects can be reliably predicted, controlled, and used in technical devices such as windows, filters, thermistors, and resistors. These effects, however, are little understood. Material of not only high resistivity but also highest purity (below the ppm limit) will be necessary to investigate the role and the nature of traps, the magnetic effect, etc. There is reasonable hope that such material would contain a much smaller number of traps and would exhibit at least some of the conventional semiconductor contact effects.

Acknowledgment

The authors wish to thank Dr. N. P. Nies of U.S. Borax Research Corp., Dr. D. R. Stern of American Potash & Chemical Corp., Mr. W. E. Medcalf of Eagle-Picher Co., and Mr. C. P. Talley of Experiment Inc., for the supply of material especially prepared for this research; Mr. W. R. Hansen of Battelle Institute and Mr. C. E. Harvey of Harvey Associates for difficult and novel analytical work.

The authors also express their gratitude to their colleagues of this laboratory who assisted in the work, especially to Mr. J. W. Mellichamp and Mr. R. O. Savage for analytical work; to Dr. J. A. Kohn and Mr. D. W. Eckart, for x-ray studies and discussions of structure; to Mrs. C. E. MacNeill for infrared data; to Mr. C. F. Cook for electron microscopy; and to Mr. H. P. Wasshausen for the painstaking ultrasonic cutting of boron crystals.

References
1. Weintraub, E., Trans. Am. Electrochem. Soc. 16 (1909) 165.
2. Weintraub, E., U.S. Patent 1,079,621.
3. Freymann, R., and Stieber, A., C. R. Acad. Sci., Paris 199 (1934) 1109.
4. Lagrenaudie, J., J. Chim. Physique 50 (1953) 629.
5. Wymelenberg, M. J., Thesis, University of St. Louis, Mo. (1958).
6. Hoard, J. L., these proceedings, p. 1.
7. Hoard, J. L., and Newkirk, A. E., J. Am. Chem. Soc. 82 (1960) 70.
8. Gaulé, G. K., Breslin, J. T., and Pastore, J. R., J. Electrochem. Soc. 105 (1958) 253 C.
9. Saratovkin, D. D., Dendritic Crystallization, translated by J.E.S. Bradley, New York, Consultants Bureau Inc. (1959).
10. Yannacakis, J., and Nies, N. P., these proceedings, p. 38.
11. Spitzer, W. G., and Kaiser, W., Phys. Rev. Letters 1 (1958) D 504 L-1-2.
12. Moss, T. S., Optical Properties of Semiconductors, London: Butterworths Sc. Publ. (1959).
13. Barnes, D., Mackenzie, R. B., and Aves, R., report R/M 125 (unclassified) of A.E.R.E., Harwell, U.K. (1957).
14. Teal, G.K., Sparks, M., and Bleuler, E., Phys. Rev. 81 (1951) 763.
15. Rose, A., "Performance of Photoconductors," Photoconductor Conf., Atlantic City (1954). See also: "Progress in Semiconductors," Vol. 2 (London: 1957, Heywood).
16. Crawford, B. L., and Edsall, J. T., J. Chem. Phys. 7 (1939) 223.
17. Price, W. C., J. Chem. Phys. 16 (1948) 894.
18. Ioffe, A. F., "Semiconductor Thermoelements and Thermoelectric Cooling," translation, London (1957).
19. Hagenlocher, A. K., thesis, Technische Hochschule Stuttgart, Germany (1958). See also: A. K. Hagenlocher, these proceedings, p. 128.
20. Medcalf, W. E., priv. comm., See also: Bean, K. E., et al., and Starks, R. J., et al., these proceedings, pp. 48 and 59, resp.
21. Greiner, E. S., and Gutowski, J. A., J. Appl. Phys. 30 (1959) 1842. See also: E. S. Greiner, these proceedings, p. 105.
22. Shaw, W. C., Hudson, D. E., and Danielson, G. C., Phys. Rev. 107 (1957) 419.
23. Becker, J.A., Green, C.B., and Pearson, G.L., Trans. El. Eng. 65 (1946) 711.
24. Sachse, H. B., Electronic Industries, Oct. 1959, p. 81.

OXIDATION OF BORON AT TEMPERATURES BETWEEN 400 AND 1300°C IN AIR

Harry F. Rizzo*

The oxidation behavior of powder compacts of both amorphous and crystalline boron was studied in air at temperatures between 400 and 1300°C. Crystalline and amorphous boron during the initial stage of oxidation followed a parabolic rate law. Crystalline boron formed a protective coating of boron oxide between 600 and 1100°C, and after 24 hours in this temperature range 16-19% of the boron was oxidized to B_2O_3. Amorphous boron oxidized at a greater rate than crystalline boron between 600 and 1000°C, and after 24 hours 22-26% of the boron was oxidized to B_2O_3. Above 1000°C vaporization of B_2O_3 took place, in addition to the formation of a brown suboxide of boron. The composition of this suboxide has been determined as B_7O. Boron nitride was also formed during the oxidation of amorphous boron between 1100 and 1300°C.

In addition to the oxidation of powder compacts of boron, a number of oxidation tests were performed with crystalline boron in the massive form. There is considerable scatter of the data for boron in the massive form which is attributed to microcracks in the boron.

The addition of silicon was found to increase substantially the oxidation resistance of boron.

Introduction

The oxidation behavior of boron was studied between 400 and 1300°C in air. The effects of particle size, type of boron (amorphous or crystalline), geometry of sample, temperature, chemical reactions, vaporization of boron oxide, and alloying elements were considered in determining the controlling mechanisms for the oxidation of boron. This study indicates that boron oxidizes to boron oxide, B_2O_3, which forms a protective layer up to 900°C. Above this temperature the flow of B_2O_3, vaporization of B_2O_3, and the formation of B_7O are all to be considered in determining the rate-controlling step. The addition of silicon improved oxidation resistance of boron remarkably, to the point where it could be used in an oxidizing atmosphere up to and possibly above 1400°C. This resistance to

---

*Aeronautical Research Laboratory, Wright-Patterson Air Force Base, Ohio.

oxidation is due, apparently, to the formation of a high-melting borosilicate glass.

Factors Affecting Oxidation

Boron has not been studied extensively until the past few years, and information in the literature is sparse and contains inconsistencies. For example, numerous text books indicate that boron oxidizes in air from room temperature to 1000°C with flame colors of red, white, or brilliant green. Talley [1], in his recent studies on the combustion of boron, considered the oxidation of boron in oxygen to be a function of both pressure and temperature; his results are discussed in this paper.

The properties of the oxidation products are of prime consideration in predicting the oxidation behavior of a material. The predominant oxidation product of boron is amorphous or vitreous boron oxide, B_2O_3. Crystalline B_2O_3, whose structure was determined by Berger [2] to be hexagonal with $a = 4.334$, $c = 8.334$ A, has a definite melting point of 450°C. Amorphous B_2O_3 is formed during the oxidation of boron, and, since it is a glass, does not have a definite melting

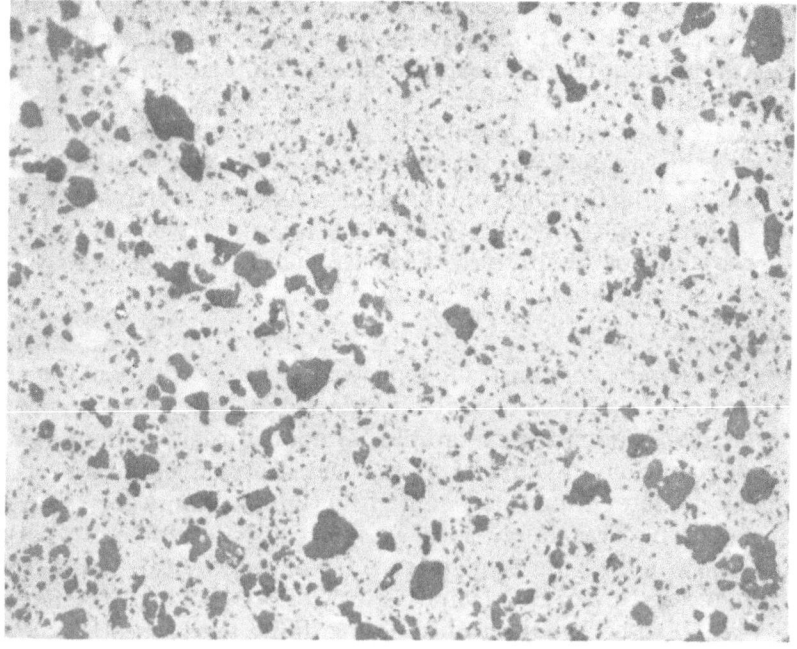

Fig. 1. Photomicrograph of a product prepared from a pellet of −325-mesh crystalline boron heated at 1300°C for 18 hours in a flowing He—O_2 atmosphere. The grey matrix phase is B_7O, the white phase is boron, and the black areas are holes which were initially filled with B_2O_3. Magnification 100×, unetched.

point. The softening temperature has been reported between 560 and 630°C. Below the temperature where B_2O_3 is too viscous to flow, or is a solid, one would expect to observe a protective coating of amorphous B_2O_3 on boron. As the temperature is increased above 630°C, B_2O_3 becomes less viscous and tends to flow under the influence of gravity which reduces the thickness of the protective B_2O_3 layer.

At approximately 1000°C and above, depending upon the pressure and presence of water vapor, the evaporation of B_2O_3 is considerable and could become the rate-limiting step in the oxidation of boron. In addition to the flow of B_2O_3 and its evaporation, a new process that has not been considered previously is the formation of the suboxide B_7O above 1100°C.

The presence of this suboxide was observed under a variety of conditions. The suboxide was formed by the reduction of B_2O_3 by Si above 1100°C and was also formed by heating mixtures of B_2O_3 and boron in air at temperatures as low as 1100°C. An oxidation product of boron above 1100°C in air was also observed to be B_7O. Figure 1 is a photomicrograph of a product prepared from a pellet of 325-mesh crystalline boron heated at 1300°C for 18 hours in a flowing He--O_2 atmosphere. The black areas are holes once occupied by B_2O_3 which was leached out during the polishing of the sample. The grey phase, which is practically continuous, is B_7O. The white phase is unreacted boron.

The suboxide has a distinctive x-ray diffraction pattern. The structure of B_7O was recently determined by Pasternak [3] to be orthorhombic with $a = 8.20$, $b = 5.35$, and $c = 5.13$ A. The Pasternak material was obtained from the American Potash and Chemical Corporation; chemical analysis indicated it to be $B_{6.8}O$ with trace impurities of magnesium and nitrogen. The density was 2.64 as compared to our determination of 2.62 g/cm^3. Pasternak relates the structure of B_7O to the tetragonal modification of elemental boron.* B_7O formed by the reduction of B_2O_3 by Si was stable at 1500°C in a vacuum, but decomposed in a vacuum at a temperature between 1500 and 1850°C. The suboxide was not soluble or attacked by either hydrofluoric or hydrochloric acids, but was almost completely soluble in nitric acid. Microhardness measurements indicated a Vickers hardness of 3600 kg/mm^2 as compared to 2950 for boron.

Since the existence of B_7O had been established, the question arose as to what effect its presence would have on the oxidation of boron. In an experiment conducted at 1000°C in air employing a powder compact of B_7O, the oxidation rate was less than one-half the

---

*Editors' note: More recent studies (Post, B., et al., Polytechnic Institute of Brooklyn, private communication) indicate that the true stoichiometry is $B_{6.5}O$ or, more properly, B_{12} (O-B-O), analogous to $B_{12}C_3$. Accordingly the indicated symmetry is rhombohedral.

rate for crystalline boron, which indicates that the presence of B_7O would reduce the rate of oxidation of boron.

Experimental Methods

The samples employed for the oxidation studies were predominantly pressed-powder pellets of both amorphous and crystalline boron. The boron was obtained from the U.S. Borax and Chemical Corporation. The crystalline boron was −325 mesh with a reported purity of 99.2%, and the amorphous boron was their 95 to 97% grade. In addition, a number of specimens were cut from melted boron samples supplied by U.S. Borax. Attempts were made to sinter boron in a vacuum, but with little success. As the melting point of boron was approached, a spontaneous reaction occurred which usually resulted in the failure of the tantalum susceptor and the formation of a number of small pieces of sintered boron.

A schematic drawing of the thermobalance employed for this study is presented in Fig. 2. A capacitance pickup at the end of the balance beam is the sensing device for the relay which provides the proper input to a reversible motor, and a potentiometer in series with the reversible motor provides the output necessary for the weight-change recorder. The samples were supported by an Al_2O_3 base and were immersed into the furnace at the test temperature. For comparative data most of the tests were extended to 24 hours.

The powdered boron was dry-pressed (small amounts of water were added to the crystalline variety to improve its pressing behavior) into discs which weighed approximately one gram each. The

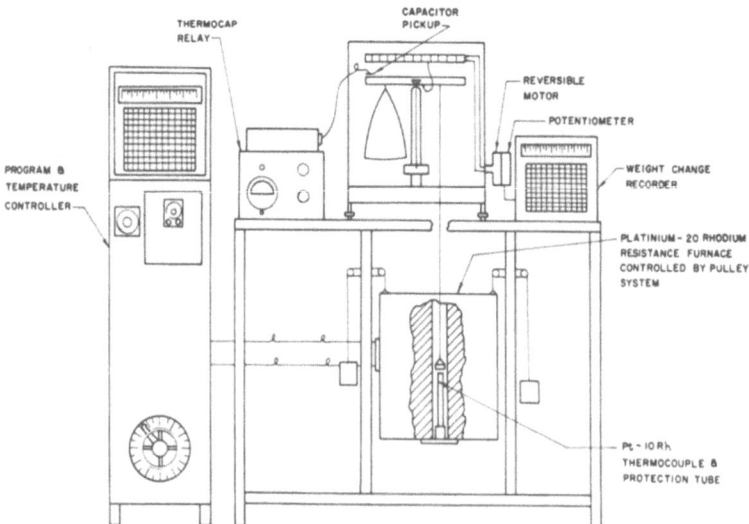

Fig. 2. Schematic of the thermobalance.

sample dimensions were taken before and after oxidation. The initial geometric surface area was employed for calculating the weight gain per unit area. Analysis of the samples after oxidation consisted of metallographic examination, surface and powder x-ray diffraction, and a determination of the amount of oxide soluble in a 40% HF solution. In addition to the amount of oxide soluble in a hydrofluoric acid solution, the boron content of the soluble oxide was determined by its precipitation as nitron tetrafluoborate [4]. The amount of B_2O_3 which flowed from the surface of a boron sample was determined by the weight gain of the Al_2O_3 holder after oxidation. Thus, by the use of weight gain data, chemical analysis, etc., a material balance should be possible which would indicate the effects of vaporization, B_2O_3 flow, and the formation of B_7O or BN. However, a material balance could not be obtained in the event of formation of B_7O or BN, since no quantitative techniques were devised to determine the B_7O or BN content. Thus, by the use of the experimental data obtainable

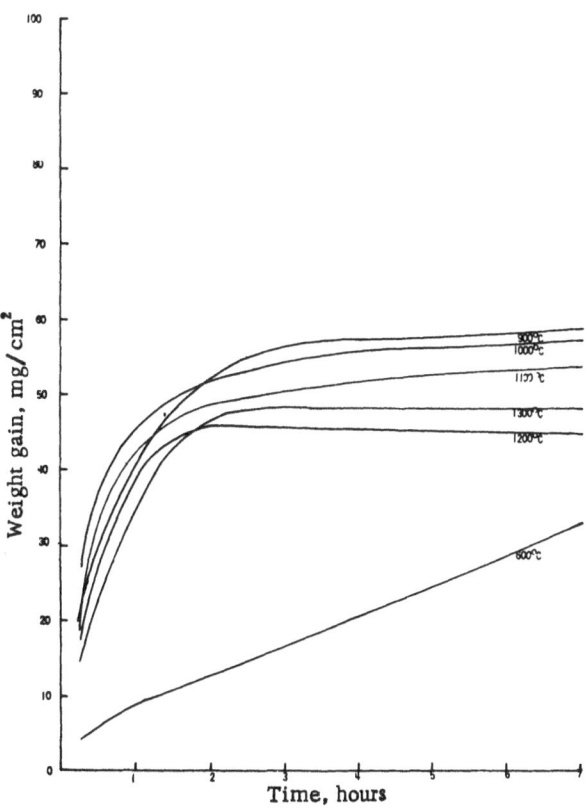

Fig. 3. Weight gain versus time curves for the oxidation of powder compacts of crystalline boron at various temperatures in air.

the oxidation of boron was studied both quantitatively and qualitatively up to 1100°C and above this temperature only qualitatively.

Results and Discussion

The oxidation data for powder compacts of crystalline boron for the first 7 hours are presented in Fig. 3. The weight gain versus time curves indicate that the oxidation of boron follows a parabolic rate law during the initial stage of oxidation.

Oxidation of the crystalline boron started at 600°C; below this temperature less than 2% of the boron was oxidized. Between 600 and 1100°C crystalline boron oxidized to B_2O_3. After 24 hours 16 to 19% of the boron was oxidized to B_2O_3. The flow of B_2O_3 from crystalline boron was negligible up to 800°C, while the vaporization of B_2O_3 was insignificant up to 1000°C. Above 1100°C the weight gain curves indicate a negative oxidation rate after the initial stages. This result may be due to either a weight loss from evaporation of B_2O_3 (which would also reduce the surface area) or the formation of B_7O, or both. As stated previously, the oxidation rate is reduced by the formation of B_7O (shown in Fig. 4 by a photomicrograph of the powdered crystalline boron after oxidation at 1300°C for 24 hours). The black holes

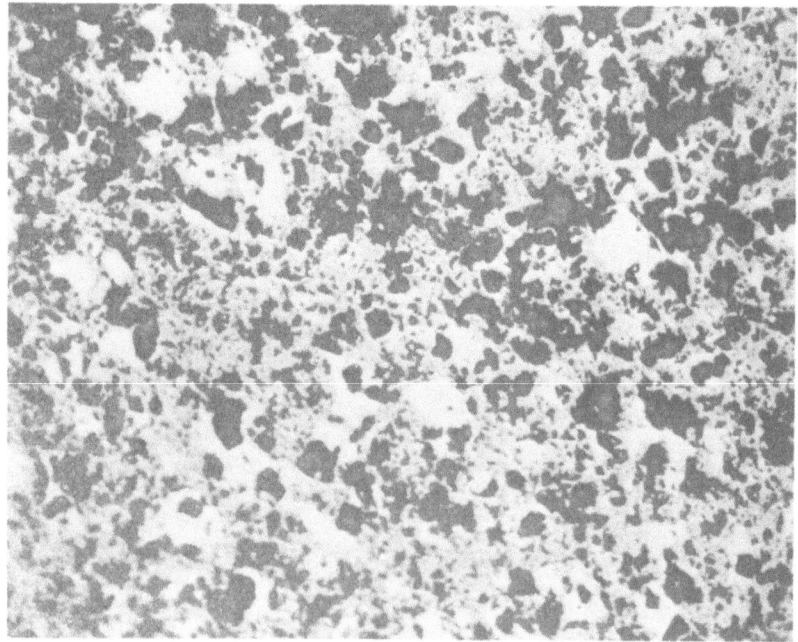

Fig. 4. Photomicrograph of a powder compact of crystalline boron after oxidation at 1300°C in air for 24 hours. The grey phase is B_7O, the white phase is boron, and the black areas are holes. Magnification 200× unetched.

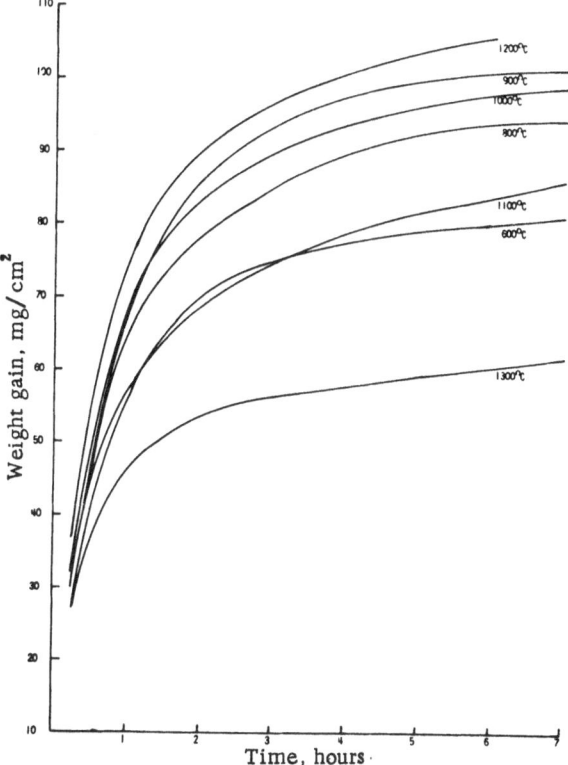

Fig. 5. Weight gain versus time curves for the oxidation of powder compacts of amorphous boron at various temperatures in air.

were initially occupied by B_2O_3, which was leached out, and the grey phase, which is the bulk of the sample, is B_7O.

The oxidation of powder compacts of amorphous boron follows a parabolic rate law during the initial stages of oxidation, as shown in Fig. 5. In addition to the formation of B_7O above 1100°C for amorphous boron, the formation of BN was confirmed at 1100°C, and above, by x-ray analysis.

Oxidation data of the amorphous boron for 24 hours are presented in Fig. 6. After 24 hours at 400°C, 6% of the boron was oxidized, while between 600 and 1000°C, 22 to 26% of the boron was oxidized. Thus in the temperature range from 600 to 1000°C the primary mechanism of oxidation is the formation of a protective coating of B_2O_3. The flow and vaporization of B_2O_3 were again negligible up to 800 and 1000°C, respectively.

The effect of particle size (or effective surface area) is demonstrated by the projection of oxidation data, shown in Fig. 6, for both powder compacts and solid crystalline boron samples. Oxidation of

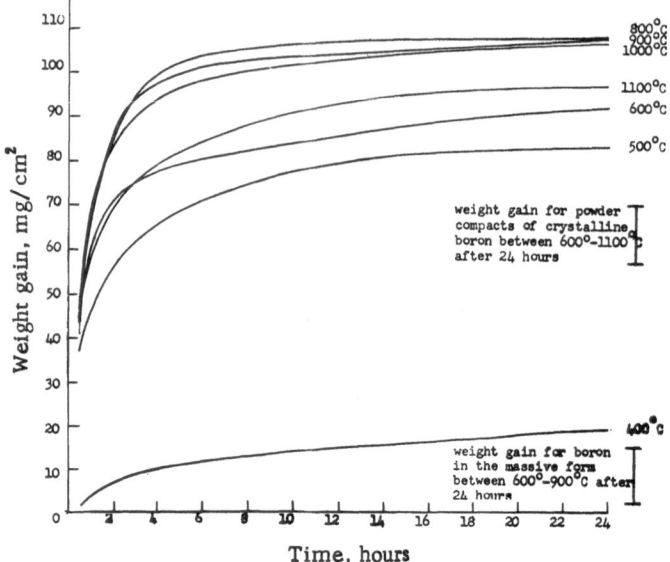

Fig. 6. Oxidation data for powder compacts of amorphous boron and projection of the oxidation data for both powder compacts and solid boron after 24 hours.

crystalline compacts after oxidation for 24 hours at temperatures between 600 and 1000°C is indicated at approximately 60 mg/cm², as compared to 105 mg/cm² for the amorphous boron. The amorphous boron has an average particle size of 0.5 to 1 μ, compared to 40 μ for the crystalline boron compacts. The projection of data for boron in the massive form up to 900°C after 24 hours of oxidation indicates a maximum weight gain of 13 mg/cm². Weight gain versus time curves were not plotted owing to the poor reproducibility caused by cracks and porosity in the samples. From the data presented it is apparent that below 1000°C boron oxidizes to form a protective coating of B_2O_3, and the amount of oxidation increases with increasing exposed surface area. The difference in the oxidation of amorphous and crystalline compacts between 600 and 1000°C is attributed to the effective surface area, since Talley [5], under similar conditions, reported identical oxidation rates for both the amorphous and crystalline boron in the massive form.

The effect of B_2O_3 flow from the surface of boron was observed by tilting the sample (massive boron oxidized at 900°C in air for 24 hours) in the Al_2O_3 holder shown in Fig. 7. The cross section of this sample is shown in Fig. 8. Rounding of the top edges of the boron sample suggests that B_2O_3, formed on the top of the specimen, flowed under the influence of gravity and allowed further oxidation of the boron. The thickness of the B_2O_3 layer varied between 40 and 200 μ.

Fig. 7. Alumina (Al_2O_3) holder containing a solid boron sample after oxidation at 900°C for 24 hours in air.

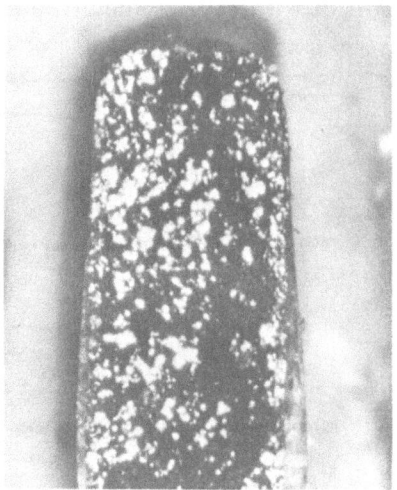

Fig. 8. Cross section of the boron sample shown in Fig. 7. The top of the sample, which was initially square, is now rounded due to effect of gravity on the melted B_2O_3 layer.

The rate of oxidation of this sample in air agrees very well with Talley's results in pure oxygen at 1 atm pressure, which indicates that the rate-determining step at this temperature is not the transport of oxygen to the surface of the boron. Since the vaporization of B_2O_3 is negligible at this temperature, the controlling mechanism is the flow of B_2O_3 from the surface of the sample, which is a function of position and geometry of the sample.

The presence of a new form of crystalline B_2O_3, recently reported by Senkovits and Hawley [6], was observed on the surface of oxidized boron samples by x-ray diffraction techniques. It is difficult to explain the presence of crystalline B_2O_3 after quenching the oxidized samples in air at temperatures as high as 1300°C. On heating the sample it was observed that the x-ray pattern, similar to that of boric acid (H_3BO_3), disappeared above 150°C and that upon cooling, apparently, there was sufficient moisture pickup to reproduce the original pattern. The reported new crystalline modification of B_2O_3 is assumed to be a hydrated form which loses water of hydration and transforms into amorphous B_2O_3 above 150°C.

A summary of the rate-limiting processes taken from Talley's studies on the oxidation of boron is presented in Fig. 9. In Region I, the rate-limiting process is the diffusion of oxygen through a glassy film, which results in a slow reaction rate. From the results of our studies the upper limit of this region should be extended to approx-

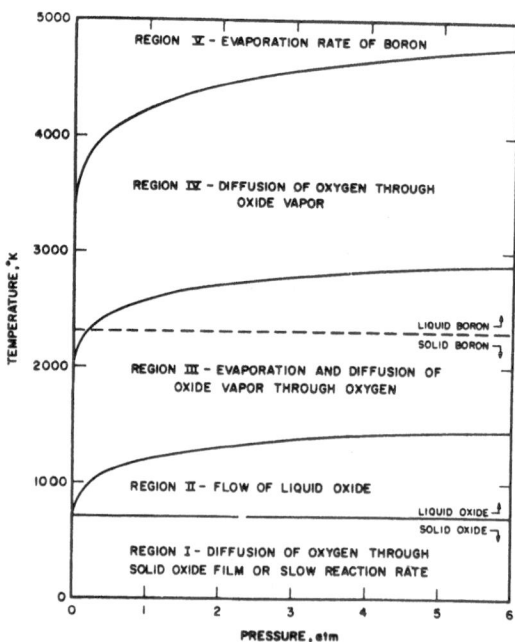

Fig. 9. Semiquantitative map showing various rate-limiting processes in the oxidation of elemental boron in oxygen as a function of temperature and pressure. After Talley [1].

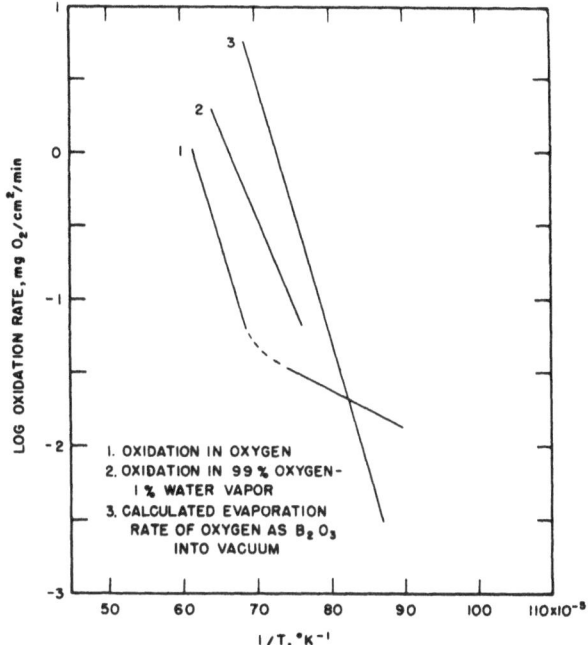

Fig. 10. Logarithm of oxidation rate versus $1/T$ for polycrystalline boron in oxygen and a 99% oxygen—1% water vapor mixture at 1 atm. After Talley [7].

imately 1000°K. In Region II, the rate-limiting process is the flow of liquid B_2O_3, and from our results the upper limit at 1 atm should be extended to 1300°K or higher. In Region III, the evaporation and diffusion of B_2O_3 vapor through oxygen are shown as the rate-limiting processes. Within this region the formation of B_7O may alter the rate-limiting process and must be considered. Regions IV and V are concerned with the combustion of boron, which is not considered here.

A summary of Talley's oxidation rates for boron is presented in Fig. 10 [7]. Curve 1 for the oxidation of boron in oxygen at 1 atm indicates two rate-limiting processes. Talley obtained an activation energy of 12 kcal/mole for the lower-temperature region, where the flow of liquid B_2O_3 is the rate-limiting process, and an activation energy of 77 kcal/mole for the high-temperature region, where evaporation of B_2O_3 is the controlling mechanism. Curve 2 indicates a high rate of oxidation (in the presence of H_2O vapor), which is probably due to the vaporization of gaseous species other than B_2O_3 [8]. The evaporation of B_2O_3 in vacuo, plotted as Curve 3, indicates that the rate-controlling process at high temperature is due to evaporation of B_2O_3.

The results of the present study indicate that boron can be utilized up to 900°C in an oxidizing atmosphere provided particular

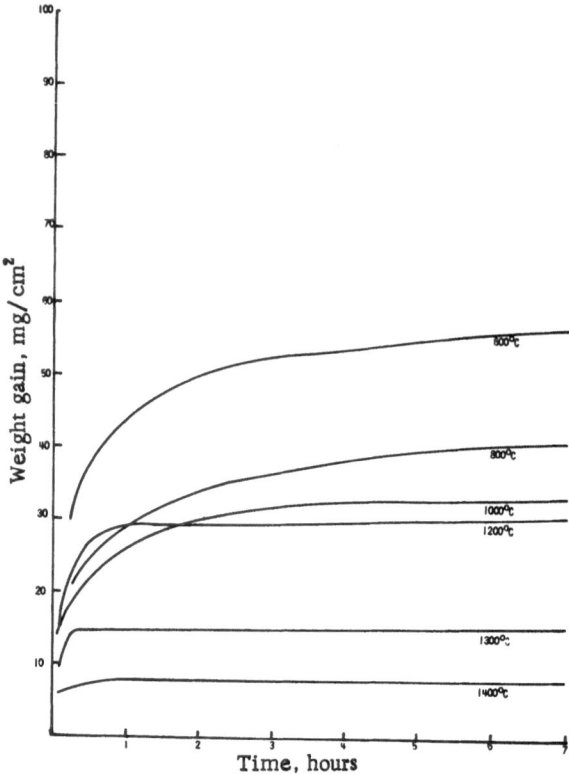

Fig. 11. Weight gain versus time curves for the oxidation of powder compacts of amorphous boron with 50 atomic percent silicon at various temperatures in air.

attention is paid to the position and geometry of the sample at the higher temperatures. One method to increase the oxidation resistance of boron would be to alloy boron with other elements. These elements should either form a protective coating of their respective oxides or react with B_2O_3 to form high-melting glasses. To be effective, these glasses must have low vapor pressures.

The addition of silicon to boron was found to increase markedly the oxidation resistance of boron [9]. The oxidation data in air for powder compacts of B with 50 atomic percent Si are shown in Fig. 11. It is to be noted that the extent of oxidation decreases with increasing temperature. The oxidation data for B with 25 atomic percent Si are presented in Fig. 12. The trend of decreased oxidation with increasing temperature again is apparent. The oxidation curves for Si and various Si--B mixtures at 1200°C are shown in Fig. 13. The oxidation of Si--B mixtures at this temperature is reduced with increasing boron concentration. It is evident that decreased oxidation occurs with both increasing temperature and increasing boron concentration.

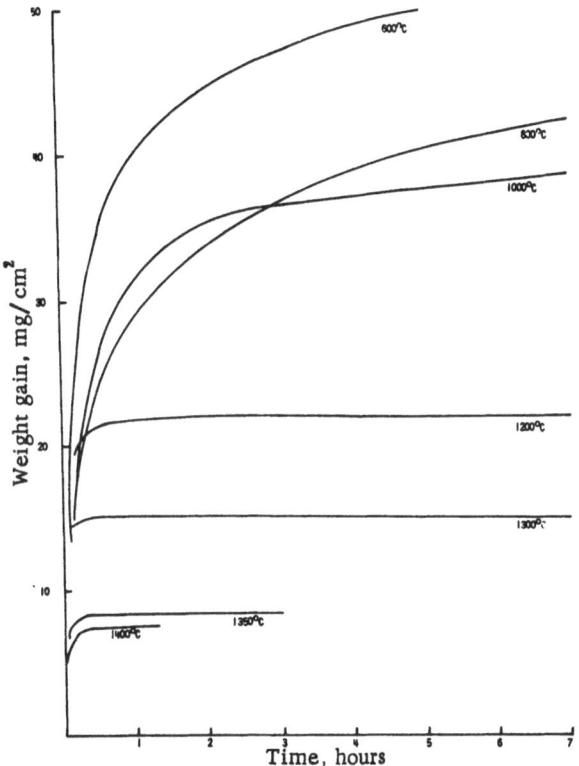

Fig. 12. Weight gain versus time curves for the oxidation of powder compacts of amorphous boron with 25 atomic percent silicon at various temperatures in air.

An explanation for the reduced oxidation is the formation of a dense borosilicate phase which retards further oxidation.

Summary

It has been shown that between 400 and 1000°C the oxidation of powder compacts of amorphous and crystalline boron follows a parabolic rate law during the initial stages of oxidation and that the basic mechanism is the oxidation of boron to B_2O_3, which forms a protective layer. Between 1000 and 1300°C the oxidation of boron is quite complex and is dependent upon the flow of liquid B_2O_3, the formation of B_7O, and the evaporation of B_2O_3.

In an effort to improve the oxidation resistance of boron it was observed that the addition of Si apparently produced a high-melting borosilicate phase which, once formed, practically eliminated further oxidation.

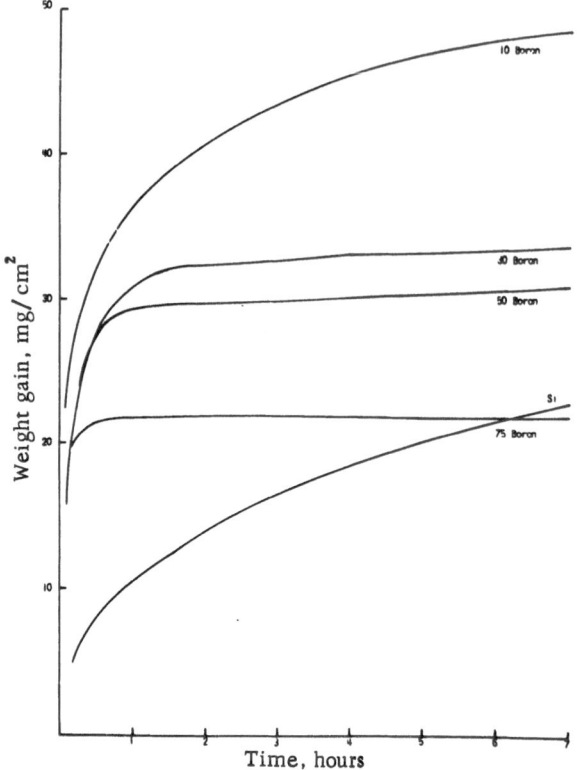

Fig. 13. Weight gain versus time curves for powder compacts of silicon and various silicon—boron (amorphous) mixtures at 1200°C in air.

Acknowledgment

This work was accomplished at the Aeronautical Research Laboratory, Wright Air Development Center. The author wishes to express his appreciation·for the Air Force's permission to publish the results of this investigation. The assistance of members of the Metallurgy and Ceramics Research Branch is gratefully acknowledged as follows: H. J. Garrett, P. R. Mason, and L. R. Bidwell for x-ray studies; F. W. Vahldiek, W. C. Simmons, H. O. Bielstein, and A. J. Stec for various experimental phases of work; and B. C. Weber for his suggestions and review of this work. The kind and helpful encouragement of E. J. Hassell is also appreciated.

References

1. Talley, C. P., "Combustion of Elemental Boron," Aero Space Eng. 18, No. 6 (1959) 37-41.
2. Berger, S. V., "The Crystal Structure of Boron and Boron Oxide," in Proceedings of the International Symposium on the Reactivity

of Solids, Part I (Goteborg: Elander Boktryckeri Aktiebolaj, 1954) pp. 422-426.

3. Pasternak, R. A., "Crystallographic Evidence for the Existence of B_7O," Acta Cryst. 12, No. 8 (1959) 612-613.

4. Lucchesi, C. A., and DeFord, D. D., "Gravimetric Determination of Boron by Precipitation as Nitron Tetrafluoborate," Anal. Chem. 29, No. 8 (1957) 1169-1171.

5. Talley, C. P., "Combustion of Boron," Experiment Incorporated TM-1009, November 1, 1957.

6. Senkovits and Hawley, ASTM Card No. 6-0297.

7. Talley, C. P., "Combustion of Boron," Experiment Incorporated TM-1094, February 1, 1959.

8. Margrave, J. L., et al., "Structures and Thermodynamic Properties of High Temperature Gaseous Species in the B_2O_3-H_2O System," University of Wisconsin, ccc-1024-TR-231, March 14, 1957.

9. Rizzo, H. F., Powell, J. F., Vahldiek, F. W., and Weber, B. C., "Oxidation of Boron and Silicon—Boron Compositions in Air at Elevated Temperatures," presented at the Annual Meeting of American Ceramic Society, Chicago, Illinois, May, 1959.